International Archives of the History of Ideas
Archives internationales d'histoire des idées

Founding Editors
Paul Dibon
Jeremy Popkin

Volume 250

International Archives of the History of Ideas/Archives internationales d'histoire des idées is a series which publishes scholarly works on the history of ideas in the widest sense of the word. It covers history of philosophy, science, political and religious thought and other areas in the domain of intellectual history. The chronological scope of the series extends from the Renaissance to the Post-Enlightenment. Founded in 1963 by R.H. Popkin and Paul Dibon, the International Archives of the History of Ideas/Archives internationales d'histoire des idées, edited by Guido Giglioni and John Christian Laursen, with assistance of Former Director Sarah Hutton, publishes, edits and translates sources that have been either unknown hitherto, or unavailable, and publishes new research in intellectual history, and new approaches within the field. The range of recent volumes in the series includes studies on skepticism, astrobiology in the early modern period, as well as translations and editions of original texts, such as the *Treatise of the Hypochondriack and Hysterick Diseases* (1730) by Bernard Mandeville. All books to be published in this Series will be fully peer-reviewed before final acceptance.

Daniel Špelda

The Origins of the Idea of Scientific Progress

Bernard de Fontenelle and His Contemporaries

Springer

Daniel Špelda
Department of Philosophy
Masaryk University
Brno, Czech Republic

ISSN 0066-6610 ISSN 2215-0307 (electronic)
International Archives of the History of Ideas Archives internationales d'histoire des idées
ISBN 978-3-031-60525-3 ISBN 978-3-031-60526-0 (eBook)
https://doi.org/10.1007/978-3-031-60526-0

This Springer imprint is published by the registered company Springer Nature Switzerland AG
The registered company address is: Gewerbestrasse 11, 6330 Cham, Switzerland

Acknowledgments

The work on this monograph was supported by the Czech Science Foundation, grant 20-01948S.

I thank the two anonymous reviewers for their constructive comments that improved the final form of the book.

I thank Eliška Peřinová for her great help in editing the manuscript. I also thank the team of editors at Springer, who were helpful and always ready to advise on finalizing the manuscript.

All translations of quotations are mine unless otherwise indicated.

Contents

About the Author

Daniel Špelda works as a professor at the Department of Philosophy in the Faculty of Arts at Masaryk University in Brno (Czech Republic). In his research, he has concentrated on the philosophical problems, concepts, and categories that played a role in the development of early modern science, including curiosity, teleology, progress. In recent years, he has addressed the question of how the early modern natural philosophers understood the history of their field and their position in the course of cultural history.

Chapter 1
The Concept of Progress

Contents

Abstract The concept of progress is often applied by philosophers and historians in confusing ways; the concept deserves clarification. First, it would be appropriate to make distinctions between the general progress of mankind, different kinds of progress in various fields of human activity (art, law, and philosophy), and the idea of scientific progress. There is a great deal of consensus in existing research that the idea of scientific progress became the model for all kinds of progress at the beginning of the Enlightenment. What is lacking, however, is a convincing answer to the question of where the idea of scientific progress came from. In this chapter, three previous interpretations of the origin of the idea of (scientific) progress are rejected: that the idea was the result of the advancement of knowledge; that the idea was the result of the secularization of eschatology; and that the idea existed in classical antiquity and was only revived in early modern times. This chapter also outlines the methodological tenets of the book. From the methodological point of view, the presented research is situated in the vicinity of Lorraine Daston's historical epistemology and subscribes to Hans Blumenberg's philosophical analyses of early modern science and philosophy. The main thesis of the book claims that the idea of scientific progress represented a methodological response to the contingency of nature. When nature was conceived as an inactive set of corpuscles that is completely indifferent to the human desire for knowledge and human cognitive capacities, it became clear that knowledge of nature must adopt a temporal regime. The bearer of this temporal regime was an institution.

D. Špelda, *The Origins of the Idea of Scientific Progress*, International Archives of the History of Ideas Archives internationales d'histoire des idées 250,
https://doi.org/10.1007/978-3-031-60526-0_1

Keywords Idea of scientific progress · Secularization · Philosophy of history · Progress of humankind · Historical epistemology

1.1 The Absence of the Idea of Scientific Progress

The traditional historiography of early modern science focused on the great figures of the so-called Scientific Revolution. Recent research has begun to explore new topics much more intensively. Historians have been examining important categories of early modern science, such as probability, fact, explanation, method, hypothesis, experiment, or observation. At the same time, methodological tenets from science studies have been transferred to the study of the history of science, and historians have carefully examined the sites of the production of scientific knowledge, its material side, the visual culture, the colonial aspects of science, the relationship of research to the state and the commercial world, the globalization of science, the *personae* of scientific research and their historically fluid identities.[1] All of these investigations have contributed greatly to transforming and deepening our understanding of early modern science, particularly in comparison to how it was interpreted by historians working between approximately 1920 and 1970.

The idea of scientific progress does not appear at all in recent scholarship on early modern science and philosophy. It does not occur in the third volume of *The Cambridge History of Science*, nor can it be found in *The Cambridge History of Seventeenth-Century Philosophy, A Companion to Early Modern Philosophy*, nor *The Cambridge History of Philosophy of the Scientific Revolution.*[2] No entry on progress appears even in the most recent and most detailed work on early modern thought, the *Encyclopedia of Early Modern Philosophy and the Sciences*.[3] All of these works mentioned traditional themes related to early modern science such as the mathematization of nature, experimental methods, and empiricism, but the belief in future advances in natural philosophy has disappeared from historical works on early modern natural philosophy, even though it was expressed repeatedly by the natural philosophers of the period.

Some historians may assume that the idea of scientific progress is self-evident and natural, posing neither historical nor explanatory problems. Others may believe that the category of scientific progress is too closely associated with the triumphalist history of science that contemporary historians wish to avoid.[4] Still other historians may have been influenced by the massive criticism that the idea of progress suffered

[1] See e.g. Daston and Park 2006; Van Damme 2015.

[2] See Garber and Michael 1998; Daston and Park 2006; Nadler 2008; Miller and Jalobeanu 2022.

[3] Wolfe and T. and Dana Jalobeanu. 2020.

[4] See e.g. index in Applebaum 2000, where 'progress' refers to the positivist historiography of science.

in continental philosophy during the second half of the twentieth century. This criticism seems to have loaded the very notion of progress with such negative connotations that it is not considered worthy as a subject of serious historical inquiry.

I believe that the idea of scientific progress is a legitimate subject of historical inquiry. The philosophical literature of the seventeenth century contains many references to 'the advancement', *les progrès*, or *progressus* of natural philosophy. The idea of progress is thus an actor's category, and it is neither necessary nor appropriate to burden it with the connotations it has obtained in contemporary philosophy, science, humanities, and developmental studies.

I want to approach the examination of the origins of the notion of scientific progress from a historical perspective without taking into account later debates about its legitimacy in twentieth-century philosophy. My work, however, does not fall within the history of ideas or intellectual history. My aim is not to pinpoint the originators of the idea if any. Nor do I wish to describe the spread of the idea in the intellectual context of the seventeenth century. Above all, I want to show that the idea of scientific progress was not self-evident in the context of European thought and that it emerged in natural philosophy as a consequence of the changes in ontology, anthropology, and epistemology.

As the concept of progress is burdened with many debates, connotations, and misunderstandings that emerged especially during the twentieth century, I want to present my understanding of the concept in this introductory chapter. In the following sections, I specify the concept of scientific progress itself; present the current state of the research; explain my approach; and describe the aims of this book.

1.2 The Idea of the General Progress of Humankind

The idea of progress was a major theme of continental philosophy in the second half of the twentieth century. Philosophers questioned the meaning, existence, benefits, and justification of progress. French and German philosophers in particular, from the Frankfurt School to the postmodernists, attacked the modernist belief in progress from various points of view. They considered it an ideological product of the Enlightenment's drive to improve the world without regard for the will of its inhabitants. Philosophers such as Karl Löwith and Theodor W. Adorno identified the belief in progress as one of the main culprits in the disasters that befell modern European culture.[5]

Regardless of this critical assessment of the idea of progress, many historians and philosophers of the twentieth century were aware that the idea of progress represented one of the most important categories of modernity. From the beginning of the twentieth century, historians therefore attempted to explain the history and origins of the idea of progress. However, philosophical and historical works often

[5] See Löwith 1963; Adorno 1964; cf. also Loewenstein 2009.

used the concept of progress in a vague way, ignoring the historically shifting semantics of the concept. In my opinion, historical research thus failed to provide a sufficiently convincing explanation of its origins.

A major problem of research to date has been the lack of distinction between concepts such as the progress of humanity, the progress of knowledge, the progress of society, and the progress of history itself. As Reinhart Koselleck and Jochen Schlobach demonstrated, there was no general idea of human progress before the end of the eighteenth century. At that time, progress was spoken of in the plural followed by a genitive (e.g. *les progrés de l'astronomie*).[6] The Enlightenment philosophers distinguished several separate kinds of progress and applied each to different areas of human activity. There was the progress of science, the progress of technology, the progress of morality, the progress of civilization, the progress of society, the progress of economics, the progress of literature, the progress of art, etc. These kinds of progress were often understood to have a tempo of their own, distinct from advances in other areas of culture. Koselleck convincingly explained that a synthesis of all these kinds of progress took place at the height of the Enlightenment. The individual kinds of progress became the universal progress of history. The term 'progress' became grammatically a collective singular expression that tied numerous experiences together into a single word. According to Koselleck, this process occurred in three steps[7]:

> First of all, the subject of progress was universalized. It no longer referred to a delimitable sphere, such as science, technology, art, etc., any of which were formerly the concrete substratum of particular progressions. Instead, the subject of progress was expanded to become an agent of the highest generality, or one with a forced claim to generality: it was a question of the progress of humanity. At first, 'humanity' was not meant as the acting but rather the referential subject, for instance in the sense of those 'hypothetical people' to which Condorcet subordinates all individual instances of progress as an intellectually constructed subject. The chosen people of the Judeo-Christian heritage become the hypostasis of progress. Soon one can also speak of the 'progress of time' and much later, of 'the progress of history'. Thus, out of the histories of individual cases of progress comes the progress of history. This is the second phase. For in the course of the universalization of our concept, subject and object switch their roles. The subjective genitive turns into the objective genitive: In the expression 'the progress of time' or 'the progress of history', progress assumes the leading role. Finally, in a third phase, this expression came to stand alone: progress became 'progress purely and simply,' a subject of itself.

Following Koselleck, other historians were also convinced that at the end of the Enlightenment Age, partial kinds of progress were united in one immense historical current labelled as the progress of humankind (or "man").[8] In this book, I refer to this narrative of advancing humanity (or history itself) towards a better future as the general idea of progress. This idea constituted a key metanarrative of the modern interpretation of world history, which is why it was most often challenged by

[6] Schlobach 2001, 1100; Koselleck 2006, 81–85, 172–173; Rohbeck 2005.

[7] Koselleck 2002, 171–174. In this paper Koselleck summarized his earlier research, see especially Koselleck 1975.

[8] Wagner 2016, 23–25; see also Rohbeck 2008, 859–860; Blumenberg 1996a, 43; Zedelmaier 2003.

twentieth-century philosophical currents that were critical of modernity. This general idea of progress, its origins and history, is not the subject of this book.

The general idea of progress was created by universalizing partial kinds of progress. But where did the partial progress of individual cultural areas come from? Among historians, there is a fairly broad consensus that the progress of science was the paradigm for partial kinds of progress.[9] According to this interpretation, the idea of progress appeared first in the empirical investigation of nature (science) and was then conveyed to other cultural and social areas. The idea of scientific progress served as a formal model for conceptualizing change and improvement in other fields of human culture such as politics, arts, moral or economy because it offered an explanatory formula for the relationship between human actions and historical time that suited the interests and goals of the Enlightenment and modernity. The process of generalization and conveying the idea of scientific progress is also not the subject of this monograph. This book investigates a different question: If the general notion of progress and the partial kinds of progress originally arose from the idea of scientific progress, where did the idea of scientific progress come from?

1.3 Existing Approaches to the Origins of the Idea of Scientific Progress

The difficulty with finding the origin of the idea of (scientific) progress is that historians often do not know what they are looking for because they confuse the general notion of progress, partial kinds of progress, and scientific progress. In the search for the origins of the unspecified and very vague idea of progress (encompassing also scientific progress), three basic narratives have emerged. According to the first, the notion of progress arose as a natural consequence of progress itself. The second narrative explains the emergence of the idea of progress as the result of the secularization of eschatology. According to the third narrative, the idea of progress was already formulated in classical antiquity and its later occurrences were only diverse adoptions of the Greek notion. The interpretations of historians of science who were active in the second half of the twentieth century can be added to these three narratives as the fourth view. In order to better understand the main argument of the book, it is necessary to consider these four narratives in more detail.

The first major works on the history of the idea of progress were written in the early twentieth century by Jules Delvaille (1910) and John Bury (1920). In his book,

[9] In different variations, the idea was formulated by various historians: Bury 1920, 4, 111; Rohbeck 1987, 31, 126–132; Pomian 1988, 54; Rapp 1992, 124; Gembicki and Reichardt 1993, 131–132; Blumenberg 1996a, 38; Schlobach 2001, 1102; Taguieff 2002, 61; Rouvillois 2010, 15–17; cf. Israel 2004.

Delvaille collected progressivist-sounding quotations from the history of European philosophical literature. Bury followed Delvaille in some respects, but he rejected the presence of the idea of progress in classical antiquity and concentrated on the period from the Renaissance to the twentieth century. Both works were important because they provided readers with lists of texts and contexts in which formulations of what both historians regarded as the notion of progress appeared.

The problem with both books was the vagueness of the category of progress, which they more or less identified with overcoming the past and expecting future improvement. Both Delvaille and Bury conflated the progress of civilization, the progress of knowledge, the progress of humanity, or the progress of morality. In their internalist reading, the idea of scientific progress was just an exemplification of the abstract idea of progress. Bury and Delvaille ultimately reduced the idea of scientific progress to a resistance to the authority of the ancient authors and a vague belief in the further perfecting of knowledge. Therefore, according to both historians, the *querelle des anciens et des modernes* represented the original context in which the idea of the progress of knowledge was formulated (Bury 1920, 98–120; Delvaille 1910, 212–221).

Both Delvaille and Bury considered key texts for the genesis of the idea of progress to be Francis Bacon's *Novum Organum* (1620), the short fragment *Préface sur le Traité du vide* (1648) by Blaise Pascal, and Bernard de Fontenelle's essay *Digression sur les Anciens et les Modernes* (1688). According to both historians, these texts reassessed the relationship of philosophy to authority and emphasized the possibility of further advancing knowledge. However, this explanation of the origin of the idea of progress remained rather vague and reductive because the idea of progress brought more to the field than a simple rejection of authority and a belief in the open-ended nature of knowledge.

This is related to another problem with the books written by Delvaille and Bury. Neither historian dealt with the idea of the progress of scientific knowledge found in seventeenth-century scientific texts. Both focused on the texts that made programmatic statements about the relationship of early modern philosophy to the authority of ancient scholarship. The emphasis on capturing the historical self-positioning of early modern philosophy prevented either historian from grasping more precisely the functions that the concept of progress performed in scientific practices themselves and the motives that initiated its introduction. They understood the idea of scientific progress (or the idea of the progress of knowledge) primarily as a category explaining history of knowledge and therefore they interpreted its origins in the framework of the philosophy of history, and not in the context of scientific knowledge and epistemic practices typical of natural philosophy.

Later in the twentieth century, other historians also placed the roots of the idea of scientific progress in the early modern rejection of the authority of Greek philosophy and the *querelle des anciens et des modernes*. The three texts by Bacon, Pascal, and Fontenelle were often considered a new way of conceptualizing the history of knowledge and therefore also the roots of the idea of scientific progress. From this perspective, historians and philosophers such as Hans Robert Jauss, Hans Blumenberg, Reinhart Koselleck, Jochen Schlobach, Friedrich Rapp, and Frédéric

Rouvillois provided valuable interpretations of these texts and the philosophical context of the *querelle*.[10]

However, these historians and philosophers also understood the idea of scientific progress as a historical category that explained the course of the history of knowledge and defined the historical self-awareness of early modern natural philosophy. Unfortunately, they did not pay any attention to the role which historical time played in epistemic practices of natural philosophy and science. Thus, while there is much agreement that the development of scientific knowledge in the seventeenth century was the original source for the birth of the general idea of progress, philosophers and historians have not addressed how the category of progress was used by the seventeenth-century representatives of natural philosophy, such as Galileo Galilei, Pierre Gassendi, Jean-Dominique Cassini, Robert Boyle, Christiaan Huygens, and Isaac Newton. I argue that research conducted from the perspective of the philosophy of history will not help to apprehend the origins of the idea of scientific progress. Instead of inquiring into the period notion of history, it is necessary to study the temporal regime that emerged in epistemic practices of early modern natural philosophy.

The second big narrative about the genesis of the general notion of progress was based on the secularization hypothesis. In the 1950s and 1960s, an interpretation became widespread among philosophers, historians, and theologians that presented the origin of the (general) idea of progress as a consequence of the secularization of Christian eschatology. In his review from 1967, W. Warren Wagar mentioned five main representatives of the secularization hypothesis (in the English-language research): Karl Löwith, Reinhold Niebuhr, John Baillie, Eric Voegelin, and Emil Brunner (Wagar 1967). It would certainly be possible to name other scholars from the continental milieu, but perhaps the most famous statement of the secularization thesis in the field of the philosophy of history was a book by the German philosopher Karl Löwith, originally published in English as *Meaning in History* (Löwith 1949, cf. Barash 1998). Löwith's book dealt with the general idea of progress in the philosophy of history but did not mention scientific progress. Löwith only later applied the secularization model to scientific progress. In one of his articles, he noted that due to the success of scientific progress, the physicist had taken over the place of the theologian and planned progress had assumed the function of providence (Löwith 1963: 34).

The secularization hypothesis was assailed by the German philosopher Blumenberg in the first edition of his *Die Legitimität der Neuzeit* (1966). In the revised edition from the 1970s, Blumenberg widened and diversified his argument. The fundamental philosophical-theological dispute between Blumenberg and Löwith (and also Carl Schmitt in the field of political philosophy) has since become a standard part of the account of how the philosophy of history has been interpreted in the twentieth century. The Blumenberg-Löwith-Schmitt debate has been

[10] Jauss 1964; Blumenberg 1986, 173–208; Schlobach 1980, 226–302; Rapp 1992, 174–176; Blumenberg 1996a, 54–67, 530–543; Rouvillois 2010, 129–165.

summarized and commented on several times and has become a research topic in its own right, so I do not address it here.[11]

What is important for our topic is that Blumenberg formulated fundamental objections to the idea of secularization as an interpretive category. After Blumenberg, other historians and philosophers developed and sharpened his arguments or formulated their own reservations about the persuasiveness of the secularization hypothesis in interpreting the origins of the general idea of progress.[12] Despite this critical assessment, the idea of the secularization of Christian beliefs remains a popular interpretive paradigm for the history of certain categories of modern thought, including the idea of progress. Robert Nisbet, for example, in his widely read *History of the Idea of Progress* (1980), superficially conflated the Christian conception of history, especially Augustine's, with the modern idea of progress, without bothering to mention the serious arguments of Blumenberg and other scholars against the persuasiveness of this construction (Nisbet 1994, 75, 127).

In this book, I do not intend to go into further discussion of secularization. I believe that secularizing interpretations of the idea of progress have been convincingly challenged and disproved. I also think that even without sophisticated philosophical, historical, and theological arguments, the secularization hypothesis should be refuted simply on the basis of a lack of convincing evidence. David Spadafora, for example, simply noted a lack of evidence for Ernest Tuveson's idea that science had secularized eschatology into the idea of progress (Spadafora 1990, 127). This is a very effective remark, because indeed it is hard to find traces of secularization in the texts of seventeenth-century natural philosophers. I also agree with Peter Harrison that secularization is an inappropriate category for capturing the hybrid relations between religion and natural knowledge in the seventeenth century because it is too crude and ideologically and historically loaded (Harrison 2019; cf. Groh 2010). I argue in this book that the idea of scientific progress had its origins in sources other than Christian eschatology, and it is therefore not subject to a secularization hypothesis.

The third major narrative about the history of the concept of progress claims that the concept originated in classical antiquity. Delvaille found the origins of progress in Lucretius (1910, 67–72). Bury rejected the presence of the idea of progress in classical antiquity (1920, 9–13). In 1967, classical scholar Ludwig Edelstein published *The Idea of Progress in Classical Antiquity*, in which he collected a large number of progressivist-sounding passages from ancient literature (Edelstein 1967, 144–154). In their reviews of Edelstein's book, both Albrecht Dihle and Eric R. Dodds argued that while many quotations in Greek literature resembled the

[11] See e.g. Wallace 1981; Schmidt-Biggemann 1987; Monod 2002, 230–279; Koselleck 2003, 185–195; Gordon 2019; Griffioen 2019.

[12] Dagen 1977, 21; Seifert 1982; Seifert 1986; Spadafora 1990, 104–132; Koselleck 2006, 165–166; Günther 1996, 157–158; Rohbeck 2005, 119; Taguieff 2002, 27–29; Rouvillois 2010, 51–57; Sommer 2014.

modern notion of progress, it was possible to find equally many quotations that denied the progress and proclaimed a decline or cycles of flourishing and decay. It is thus impossible to state unequivocally whether the idea of progress existed in classical antiquity.[13]

The debate about the existence of the idea of progress in classical antiquity continued in the years following the publication of Edelstein's book. The main problem, of course, has always been the vagueness of the category of progress and the willingness of some historians to infer an intellectual affinity from the verbal analogies between ancient and modern quotations. In the 2010 edition of the classic Pauly's *Encyclopaedia of the Ancient World* (2010), Wilfried Nippel indicated in the entry 'Progress' that Edelstein's attempt to document the presence of the modern idea of progress in classical antiquity had failed. In fact, in classical Greek and Roman literature, it is only possible to find equivalents of some aspects of the modern idea of progress, such as the improvement of political organization, the succession of stages of subsistence, or the outlines of future advances in natural philosophy and sciences. Referring to the works of Eric R. Dodds (1978) and Christian Meier,[14] Nippel concluded that despite the existence of partial equivalents, a unifying general theory of progress as we know it from the modern philosophy of history cannot be found in classical antiquity.[15]

The view that the Greeks and the Romans developed only partial equivalents of the modern idea of progress can be substantiated in several ways. For example, Koselleck noted that the Greek idea of progress was different from the modern one because it was concerned with the past but did not extend to the future (Koselleck 2002, 221–222):

> In those places where progress was registered in antiquity, it always concerned a look back, not an opening up of new horizons. [...] From both past history and the comparison with contemporaneously living barbarians, we thus find, formulated in modern terms, a relative model of progress that recognizes the uniqueness and singularity of the level of civilization reached by the Hellenes. But the path does not lead to the future.

This general remark may also apply to the area where we find the nearest equivalent to the modern idea of scientific progress, which is the field of technology and empirical science. In the Greek sources, the opinion of the superiority of contemporary knowledge and the belief in the gradual improvement of technology and scientific knowledge occurred. But even in these fields, the Greek notion of progress differed from the modern one.

The most important difference was that the Greek belief in the advancement of scientific and technical knowledge had teleological features. Leonid Zhmud noted that for the Greeks 'the notion of progress of *techne* in the past and its nearing perfection in the present come together.' Zhmud called this view teleological

[13] See Dihle 1969; Dodds 1968; cf. also Kehl and Marrou 1978, 749–751.

[14] Meier 1975; Meier 1978; cf. also Thraede 1962; Thraede 1972.

[15] Nippel 1998, cf. similarly Dihle 1988; Rapp 1992, 110–114; Burkert 1997; Neuser 2011; Ritter 2021.

progressivism: 'The more striking the progress that had already been made, the more natural it seemed to believe that the efforts of contemporaries, including their own, would soon reach a perfection not, or unlikely, to be surpassed in the future'.[16] According to Zhmud, furthermore, the idea of progress was limited by rather narrow bounds. Most authors of the Classical and Hellenistic periods who were interested in the historical transformations of knowledge 'regarded progress as either already accomplished or to be completed by the generation to come.'[17] The teleological concept of progress thus provided an optimistic vision of the perfection of knowledge in the present or immediate future. But as with the concept of general progress, the path did not lead to the future.

Another argument against the presence of the idea of general progress and scientific progress in classical antiquity concerns the nature of Greek philosophy. Greek philosophy (including natural philosophy) had in many ways different functions and goals than early modern philosophy. According to Pierre Hadot, Greek philosophy was above all a means of self-creation and self-cultivation. Philosophy was a way of living, not an accumulation of knowledge, nor a means of opening the way to the future. In Hadot's view, knowledge of the natural world was only part of the formation of the self and the path to happiness. Knowledge of nature was not an end in itself and had only a limited function in the larger project of individual perfection. The therapeutic function of philosophy thus excluded concern for the progress of knowledge and happiness of future generations.[18]

All three narratives presented the history of the concept of scientific progress as part of a larger story about the history of the general concept of progress. From this followed the problems that I summarize here in a few points. The notion of scientific progress was treated as just one exemplification of a general idea, as if there were a contextually indifferent, abstract, ahistorical idea of progress that was merely applied to different domains, such as science, technology, society, agriculture, or humanity. Existing research has also suffered from reducing the idea of scientific progress to a historical category that interprets the past or the future and thus belongs to the philosophy of history. This is also related to the fact that historians and philosophers, although they have written about scientific progress, have rarely worked with the texts of representatives of early modern natural philosophy or science. A final problem is that previous research relied on analogies rather than precise contextual analyses and conceptual distinctions. Historians took isolated progressivist-sounding quotations from various texts without more carefully examining whether these quotes expressed the same progressivist paradigm of knowledge.

The last narrative of the origins of the idea of scientific progress can be found in the historiography of science. Among historians of science, there is only one clearly defined approach. This approach originated in Marxism, and its most famous

[16] Zhmud 2006, 79, 114, 121, 210–213; Meier 1978, 291, 297; Burkert 1997, 33–34; Rapp 1992, 114.

[17] Zhmud 2006, 60; similarly Lloyd 1999, 330, note 147.

[18] Hadot 2002, 209–211, 220–231; Hadot 1995, 63–97, cf. Dodds 1978, 24; Rapp 1992, 104–106.

example was an article by the Marxist historian and sociologist Edgar Zilsel (1945). According to this interpretation, the idea of progress did not exist in classical antiquity and came to science from outside, especially from the field of artisanal labour during the rise of capitalism in the early seventeenth century.

Zilsel's sociologizing interpretation of the idea of progress was part of a more general conception that came to be called the Zilsel thesis. According to this thesis, modern science had its origins in early capitalism, the development of technologies, and the interaction of artisans with theorists (Krohn 1976).

In the 1960s and 1970s, with a greater or lesser degree of continuity with Zilsel, Blumenberg, Arthur C. Keller, and Paolo Rossi emphasized the importance of the development of technology for the constitution of the idea of scientific progress.[19] I believe that this type of research is built on outdated methodological foundations and purposely chose cases from the history of early modern science in which the connection between scientific theory and technical practice was evident. Early modern scientists, however, often expressed the notion of scientific progress without any connection to technology or to the practice of artisans. In the following chapters, I argue that epistemology, rather than technical improvement, was the dominant context for the emergence of the idea of scientific progress.

For Zilsel, Keller, and Rossi, the aforementioned problems of three narratives also apply. These historians of science also chose progressivist-sounding quotations from Renaissance and early modern literature focusing on authors such as Loys Le Roy or Fontenelle. From these quotations, they then constructed the idea of scientific or technological progress without taking into account the intellectual context in which these quotations were formulated.

The ordering of progressivist-sounding quotations without considering their context appears in other articles on the history of the idea of scientific progress. For example, both Alistair C. Crombie (1975) and Plinio Prioreschi (2002) in their articles assumed a kind of timeless and abstract idea of progress that is in turn exemplified in history by different authors. I do not think that making lists of various statements that resemble a later idea of scientific progress is the right way to understand the origin of this idea. It must be understood as part of concrete epistemic practices.

1.4 The Idea of Scientific Progress as an Epistemic Category

The definition of scientific progress in the contemporary philosophy of science does not help our inquiry much. Definitions of scientific progress are analytical and already focus on the specific problems of scientific change in contemporary or recent science, which is different from early modern natural philosophy. Philosophers of

[19] Keller 1950; Keller 1972; Blumenberg 2013; Rossi 1996; Rossi 1970; Blumenberg 2009, 18–37.

science also follow agenda that differs from the theme of this book. They argue in favour of various models of scientific progress and examine whether progress is a normative or descriptive category, etc.[20] They also often seem to take the idea of scientific progress for granted without questioning its origins. The main problem, however, is that the contemporary philosophy of science presupposes the timeless existence of certain features of the idea of progress such as the temporal regime of acquiring knowledge and the collectivity of epistemic practices. These epistemic standpoints are neither natural nor obvious and have come to science and natural philosophy under certain circumstances, some of which I want to clarify in the following chapters.

I would characterize the idea of scientific progress quite generally as an idea that considers time a relevant epistemic factor in the knowledge of nature. This is not trivial, because until the end of the Renaissance the tradition of natural philosophy believed that knowledge of nature did not depend on time, or more precisely, time was not considered an important condition for producing knowledge of nature. Introducing time as a relevant epistemic factor meant a complex reconfiguration of ways of knowing nature that involved the emergence of new epistemic practices and the birth of new epistemic virtues that significantly transformed natural philosophy. Thus, I understand the idea of scientific progress, with all its connotations and implications, primarily as part of a new type of epistemology which included the following three features: the temporal regime of acquiring knowledge, the collectivization of research, and the necessity of progressive development in distinct phases.

In order to better understand this epistemological reconfiguration I propose distinguishing two slightly different ideas of scientific progress: descriptive and prescriptive.[21]

The descriptive idea of scientific progress arose from looking into the past and comparing it with the present. On the basis of this comparison, the concept conveys the superiority of existing knowledge over previous knowledge. Knowledge is always perfectible and has a cumulative nature. It improves and grows in proportion to the passage of time. Analogous to biological maturation processes, later knowledge is always more mature. Analogous to the mechanical processes of accumulation, later knowledge is always more extensive because the course of time provides more experience. Thus, descriptive progress is primarily an expression of the experience of accumulating knowledge.

The prescriptive idea of progress encompasses the descriptive notion, but it also includes something more: It is future-oriented and provides natural philosophy with an open-ended character. It is an epistemic category that contributes to the expansion of knowledge and prevents the closure of knowledge into a definitive system. The prescriptive idea of progress encompasses epistemic attitudes such as long-term data

[20] Losee 2004; Niiniluoto 2019.

[21] Both concepts are mentioned in the context of progress by A. C. Crombie, without explaining and elaborating on them; see Crombie 1997, 48.

gathering, constant perfecting of methods and instruments, collective collaboration, data sharing, postponing conclusions to the future, and rejecting preconceptions. It thus brings to natural philosophy a relativist belief that every moment in this development is only temporary because every piece of knowledge will be surpassed or even denied by unpredictable discoveries in future. Each historical form of natural philosophy corresponds to currently available data and the ability to cope with error and prejudice.

Based on comparison with the past, the descriptive idea of scientific progress of knowledge led to the conclusion that the current generation knew more than its predecessors because it had data over a longer period and did not have to start from scratch, i.e. it could build on the work of its predecessors. The prescriptive idea of progress moreover involved the notion that existing knowledge becomes a preparatory stage for the more perfect knowledge of future generations. Contemporary natural philosophers are the precursors of their successors.

The descriptive idea of progress essentially articulated the relation of present knowledge to the knowledge of the past emphasizing the superiority of present knowledge over that of predecessors. We can find the idea in classical antiquity in the form of teleological progressivism. This idea also appeared in all the historical forms of the quarrel between the Ancients and the Moderns. As August Buck, Hans Baron, and Charles Trinkaus[22] have shown, the idea that I labelled the descriptive idea of progress can be found without difficulty in Renaissance versions of the dispute between the Ancients and the Moderns, where it served as an argument against the slavish worship of classical antiquity.

In the early modern period, natural philosophy introduced the prescriptive idea of progress as its essential component. Thus, the progress of knowledge no longer meant a mere assertion of superiority over predecessors. It also assumed the necessity of further development, the open-ended character of knowledge, and the historical relativity of sciences.

It is the prescriptive idea of the progress of knowledge that I understand as the proper idea of scientific progress. My book focuses on the origin and legitimation of this idea.

I would like to finish this section with a minor terminological note. The term 'scientific progress' appears in the title of this book, and for the most part in its text. The term is an admitted anachronism. The contemporary history of science, in its recent methodological self-reflection, has shown how tricky it is to carry the modern concept of science and its derivatives into the past. The ancient, medieval, and early modern disciplines of *artes liberales* (*quadrivium*), *naturalis historia*, *philosophia naturalis*, and *scientiae mixtae* should not be unwisely, ahistorically, and simplistically confused with science and its connotations in the modern sense. Yet it seems practical to me to continue to use the term scientific progress. Replacing it with equivalents that would precisely fit the historical context would require complicated formulations that would greatly burden the reading of this book. Even though I retain

[22] Buck 1958; Baron 1959; Trinkaus 1987.

the traditional term scientific progress in my text, I try to avoid the term science (unless it is used as an actor's category) and use the term natural philosophy. For other branches of knowledge that did not fall under natural philosophy, I use their specific names, such as natural history, mixed sciences, astronomy, or optics.

1.5 Methodological Background

This book is a work in the history of natural philosophy and thus stands at the border between the history of philosophy and history of science. Historical research interpreting early modern natural philosophy has in recent years focused quite significantly on the non-epistemic context. The history of science has abandoned the internalist paradigm established by Herbert Butterfield and by Alexandre Koyré and has begun to describe the social and cultural aspects of the history of science by adopting inspirational moments from the sociology of science and science studies. This practical turn in the history of science has undoubtedly broadened the knowledge of the history of modern science and greatly enriched research with new topics and approaches.

In a recent article, however, Tamás Demeter noted that since the publication of the famous book *Leviathan and the Air-Pump* (1985) by Steven Shapin and Simon Schaffer, epistemological stances and debates have been attributed to clashes of political and theological ideologies. Demeter called for a change: 'with an awareness of (. . .) context, it is now worth turning back to questions of the epistemic content itself' (Tamás 2015, 2). I share this opinion. The current social history of science tends to neglect the intellectual context and epistemological nature of some of the problems of early modern science.

Dmitri Levitin has come out even more strongly against the social history of early modern science, writing that the obsession of social historians of science with controversy, politics, and power led to accounts that are often parochial, i.e., concentrating on British culture and society, but still formulating general conclusions. In contrast, Levitin justly noted two important things. First, early modern natural philosophers understood their theories primarily as philosophy, just as their Renaissance, Medieval, and Greek predecessors did. We should therefore also understand and interpret their work as philosophy. Second, the continental natural philosophers conducted natural philosophical research in a way similar to their British colleagues. The research at the Académie des sciences in Paris was much closer to the British experimental philosophy than present-day social historians assume.[23]

I agree with both of Levitin's points. The social history of science has convincingly shown that some epistemological problems had their origins and solutions in

[23] Levitin 2014. See also Levitin's book *Ancient Wisdom in the Age of the New Science: Histories of Philosophy in England*, c. 1640–1700 where he presents his continuity thesis (Levitin 2015).

social practices. But sometimes the social history of science provides unconvincing and contrived attempts at social interpretations of early modern sciences that obscure the simple fact that epistemological problems often had their origin in epistemology and were solved by epistemological means. I think it is right to stick to the methodological principle of parsimony and always look for proximate causes and explanations. I do not deny that epistemological problems may have their cause in social conflicts, but I assume that the search for the causes of epistemological problems should begin in epistemology, and only when such a search is unsuccessful or unconvincing should one proceed to the social context. I do not deny the importance of the social and cultural context for the development of the idea of scientific progress, but at the same time I understand this idea primarily as an epistemological category. I assert that the idea of scientific progress was meant to solve methodological and epistemological problems that arose as a result of the reconceptualization of nature in early modern natural philosophy. I leave aside its relation to the values of courtly culture, the book market, or non-European cultural areas.

My approach had two main sources of inspiration. The first was the work of Hans Blumenberg. The German philosopher can be classified as a member of the (post)-phenomenological current of continental philosophy, and his voluminous work has been studied closely in recent decades.[24] Most interpreters have focused on his phenomenological, literary, and anthropological works. Great attention has been paid to Blumenberg's grand narratives on the history of European thought, which he presented in *Die Legitimität der Neuzeit* and *Die Genesis der kopernikanischen Welt* (1975). Although both of these books were translated into English in the 1980s, they have received almost no consideration in Anglophone research on early modern philosophy and science. This is apparently for two reasons: Blumenberg presented a distinctly philosophical and internalist reading of the history of science, and his writings are rather difficult and deliberately reader-unfriendly. The first feature I openly subscribe to; the second I try to avoid.

Some of Blumenberg's grand narratives, such as the second overcoming of gnosis, have lost their appeal for contemporary scholars and have been outright dismissed as incorrect (Goldstein 1998). Others have lost their persuasiveness because of the general distrust of such all-encompassing narratives in contemporary scholarship. I share this distrust and I have no intention of imitating Blumenberg in this respect.

I believe, however, that Blumenberg's thick books contain passages in which readers can find thoughtful and insightful interpretations of texts from the history of natural philosophy and science. Blumenberg was a close reader of sixteenth- and seventeenth-century writings and was able to spot motifs in the texts that have been otherwise rarely considered in historical research. He was also able to trace the long history of philosophical metaphors and similes (such as the cave, the book of nature,

[24] See for example Flasch 2017; Zill 2020.

and lions)[25] and to make the occasional use of these metaphors by early modern scientists a link in a long chain of conceptual history.

In this respect, I find Blumenberg's historical interpretations inspiring. In addition to his aforementioned books, some chapters of Blumenberg's later book *Lebenszeit und Weltzeit* (1986) inspired the fifth chapter of this book. Following Blumenberg, I assert here that the idea of scientific progress represented (among other things) an epistemological solution to the problem of the discrepancy between the short time of individual life and the immeasurable time of nature.

Methodologically, my work was strongly inspired by historical epistemology as conceived by Lorraine Daston.[26] In her papers and books, Daston explained that concepts that were commonly considered to be a natural and atemporal part of scientific knowledge had an origin and a history. Exploring such epistemic categories as objectivity, fact, observation, curiosity, probability, evidence, and collective empiricism,[27] Daston demonstrated that they could not be understood as ahistorical entities permanently associated with an universalist and essentialist concept of science. In fact, according to Daston, historical forms of thought can be reconstructed retrospectively in which these epistemic, scientific, and methodological categories did not exist. Historical research can explain the origins of these concepts and describe how they have changed over time and how they have been transferred to domains that were originally alien to them.

Daston indicated that at the beginning of their history, objectivity, observation, curiosity, scientific laws, and other epistemic concepts had motivations and justifications that were different from how modern scientific methodology explains and legitimizes them. This is well illustrated, for example, by Daston's and Katharine Park's book on the emergence of the modern category of scientific fact, which the authors surprisingly linked to the category of the marvelous (Daston and Park 2001, 218–300). In sum, Daston sought to transform abstract and supposedly timeless scientific concepts such as facts, evidence, and objectivity 'into concrete modes of thinking, feeling, and acting that are themselves historical achievements' (Daston 2003a, 9).

Daston's historical epistemology assailed the implicit and unconquered Platonism of European thought, which led to an ahistorical hypostasis of epistemic categories. Following recent methodological developments in the history of science, historical epistemology has aimed to investigate the historical dimension of scientific knowledge, the forms of representation of knowledge, its basic categories and means of dissemination, and the practices and cultural, social, and economic networks in which knowledge was embedded. In this respect, historical epistemology is undoubtedly close to the social history of science.

More important for my research, however, was another trait of historical epistemology that Daston herself referred to as 'historicized Kantianism'. Historical

---

[25] Blumenberg 1983, 1996b, 2001.

[26] On the various forms of historical epistemology, see Feest and Sturm 2011, Rheinberger 2008.

[27] Daston 1991, 1994, 2001, 2002, 2003a, b, 2004; Daston and Galison 2007; Daston 2011.

epistemology assumes that the categories that organize our knowledge determine what can be known and how humans should grasp it through their thinking. The content of knowledge, therefore, depends on the apparatus of the categories that determine it, and this apparatus is historically mutable. Thus, Daston explained that her approach to the history of science was inspired by Neo-Kantianism. She, therefore, has presented her research as seeking the conditions that make certain ways of knowing possible. In the Kantian vein, Daston has explored the history of the categories by which our thought and experience are structured, by which our processes of reasoning and demonstration are guided, and by which our standards of explanation are justified (Daston 2003a, 15–16).

> Historical epistemology draws on the history of science and culture to provide examples of these categories, but it asks different questions: Not about how mathematical demonstrations were conducted using infinitesimals in, say, the sixteenth and seventeenth centuries, but about the history of the forms and norms of demonstrations; not about how a particular empirical discovery came about in the laboratory or in fieldwork, but about the history of fact as a modality of the empirical; not about when objectivity can be said to exist in a particular discipline, but about the history of objectivity as an ideal and a practice.

Historical epistemology thus does not look for the originators, nor for the mutual influence of different scholars. Rather, it asks the Kantian question: What were the conditions of possibility of a category of scientific thought? A similar Kantian position, joined with Michel Foucault's discursive analysis, was also taken by Ian Hacking in his book on the history of probability. Hacking too was primarily concerned with the 'preconditions for the emergence of probability' (Hacking 2006, 52). By this he meant that the historian attempts to describe the type of discourse that made certain statements possible and convincing, not to trace the transmission of an idea, as intellectual history does.[28] As Omar Nasim summarized, historical epistemology is concerned with revealing how theories, statements, and practices were made possible by certain normative regimes (Nasim 2013, 142).

In my research into the origins of the idea of scientific progress, I try to emulate what Daston has done for other supposedly timeless categories of scientific research. My research is based on the conviction that the idea of scientific progress does not belong to the categories of the philosophy of history. It is an epistemological category that helped organize scientific knowledge and co-determined its content and focus. Therefore, it can be studied in a similar way to other concepts and categories explored by historical epistemology. This approach means that my research differs from previous investigations into the origins of the idea of scientific progress. First, I do not construct a teleological account of the progress of the idea of progress that narrates a story of a gradual path from naive ideas to a developed idea of scientific progress.

Second, I am not interested in individual historical figures. I focus on the discourses and their regularities that determined the possibility of formulating statements of a certain type. I want to examine the conditions of the possibility of

[28] On this kind of research, see also Poovey 1998, 1–26.

the concept of scientific progress. I share the position of Hacking, who in his examination of the emergence of probability wrote: 'We are looking for neither precursors nor anticipations of our ideas. The preconditions for probability will consist in something that is not probability but which was, through something like a mutation, transformed into probability' (Hacking 2006, 52–53). Nor I am looking for precursors, originators, or successors. I am not interested in the occult quality of influence. Although the key figure is Bernard de Fontenelle, I do not aim to write an overview of his life, works, and social environment. I understand Fontenelle primarily as the author of statements that aptly articulated the contemporary discourse on the nature of scientific knowledge.

Third, I believe that it is important to reconstruct intellectual contexts in which the idea of scientific progress did not exist in order to reveal its conditions of possibility. In the individual chapters of my book, I demonstrate that it is possible to present intellectual paradigms in which some elements of the idea of scientific progress were not thinkable. The absence of the idea of progress is illustrated primarily by examples from three specific Renaissance discourses: Platonic philosophy, astronomy, and natural history. I hope that in this way I can better explain the deep epistemological change that brought about the emergence of the idea of scientific progress.

1.6 Aims and Scope of the Book

The focus and content of this book are based on the presented methodological reflections. The descriptive idea of progress was a relatively simple empirical finding, and its existence can be documented in classical antiquity, the Middle Ages, and the Renaissance. In my monograph I aim to explain the philosophical circumstances that created the conditions of possibility for the prescriptive idea of progress in the seventeenth century.

In the early modern period, some of the assumptions that were typical of ancient and Renaissance natural philosophy were abandoned. In their place, new presumptions were adopted that created the conditions of possibility for the vindication of the idea of scientific progress. These assumptions concerned the subject of knowledge and thus involved different anthropological, cognitive, and moral definitions of the *persona* of the natural philosopher. Reconfiguration, however, concerned both the subject and the object of knowledge, that is nature. In Daston's apt abridgement, between the Renaissance and the Enlightenment, 'nature was transformed into a different entity' (Daston 2015, 183). I want to demonstrate that reconfigurations of subject and object of natural philosophy forced the introduction of new practices and epistemic categories—and one of these was the idea of scientific progress.

In its essence, the main thesis of this book argues that the idea of scientific progress compensated for the loss of anthropocentric teleology. The prescriptive idea of progress was born at the moment when natural philosophy began to grasp nature as indifferent to whether it was investigated by humans. The anthropocentric teleology of nature assumed that nature knew of human beings searching for knowledge. The philosophical admiration of God's work was even considered an important part of the history of the Creation. In the seventeenth century, the anthropocentric view of nature was replaced by an emphasis on the contingency of nature. This reconfiguration naturally affected the epistemology of natural philosophy. As the anthropocentric concept of nature promised humans causal knowledge of natural phenomena, the rejection of anthropocentrism meant the end of this optimism. Many early modern natural philosophers shared a conviction that the knowledge of causes and essences of natural objects and phenomena was not destined for humans. The prescriptive idea of progress helped to compensate for this scepticism by turning the ideal of true knowledge of causes into a process of continuous perfection open to future changes.

In the following chapters, I concentrate on several conditions of the possibility of the idea of scientific progress: rejection of anthropocentric teleology as an infantile prejudice originating at the beginning of history; the idea of contingent and immense universe, which replaced a teleological view of nature; the introduction of temporal regime of knowledge; the institutionalization of scientific research; and joining the temporal regime of knowledge with the philosophy of history.

The key factor for the emergence of the conditions of possibility was the idea of contingency. Three forms of contingency are relevant to my interpretation. The *ontological* contingency meant that that nature was a collection of passive particles of matter that did not realize any *telos*. The second form of contingency was the *epistemological* one, which denied any relationship between human cognitive abilities (primarily the senses) and nature. The third form can be labeled as *anthropological* contingency, I use this term as a designation for the incommensurability of human life with the vastness and immensity of nature. I present the three forms of contingency in the second, third, and fifth chapters.

The prescriptive idea of scientific progress enabled natural philosophers to think perfectioning of knowledge by means of three epistemic components. The first is the temporal regime of knowledge, which included epistemological features such as patient and slow gathering data, avoiding hasty speculations, gradual development, historical relativity of knowledge, and openness to empirical invalidation of theories. The second component is the collectivization of knowledge. Only organized research within an institution could cope with immensity and variety of contingent nature. The third component was the necessity of progress. The true progress of knowledge could not be random because otherwise it could be reversible. The development of natural philosophy thus came to be imagined as a necessary, gradual, and comprehensible process that resembled the epistemological maturation of the individual from childhood naivete to a rational and mature worldview. It was this last component that also later allowed the scheme of progress to be conveyed to other areas of culture and the whole of human history. I introduce these components in the fourth, fifth, and sixth chapters.

Ian Hacking wrote that the story of the emergence of probability came to an end with the publication of Jacob Bernoulli's *Ars conjectandi* (1713) (Hacking 2006, 261). Similarly, the story of the emergence of the idea of scientific progress could be said to come to a close at the turn of the seventeenth and eighteenth centuries in the context of the activities of Fontenelle and his contemporaries working at the Académie of sciences. Fontenelle was able to synthetize the values and ideals of scientific research of his time in a coherent theory of the progress of sciences. In the 1680s, Fontenelle wrote influential philosophical writings on the history of knowledge and sciences in which he reflected on the rejection of anthropocentric teleology, highlighted the contingency of nature, and formulated the basic tenets of his theory of progress using the concept of the human spirit.

The specific content of my book is based on the outlined thesis. In the second chapter, I discuss the anthropocentric teleology of nature in selected currents of Renaissance thought. Then I outline how mechanistic natural philosophy rejected this anthropocentrism and replaced it with a vision of a plurality of worlds in an immeasurable universe of inert particles dependent on God and his decisions. The third chapter begins with an outline of the epistemological optimism of the Aristotelian tradition concerning human vision. I then explain how this optimistic assumption collapsed after the discovery of telescopes and microscopes. In the fourth chapter, I discuss the key features of the temporal regime of knowledge as it appeared in the works of academicians at the Académie des sciences in Paris.

The fifth chapter discusses how the bearer of natural philosophy changed. Renaissance natural philosophers, natural historians, and astronomers still believed that knowledge of nature was an essentially individual and private matter. The research carried out by the Académie demonstrated the necessity of a collective subject, that is, an institution with a long memory and ambitions that went beyond the horizon of a single lifetime.

In the last chapter, I am concerned with connecting the temporal regime of scientific knowledge with history by means of the category of the human spirit. While Renaissance authors still mostly believed in cycles of intellectual flourishing and decline, Fontenelle and his colleagues were already working with the idea of a gradual, logical, and emancipatory process to which human knowledge was subjected through history. I do not claim that Fontenelle was the originator of the idea of scientific progress. I only believe that his texts reveal a mature synthesis of the various components of the prescriptive idea of progress as they appeared in various discourses of his time: cosmological questions, relation to ancient philosophy, the rhetoric of utility, accelerating the development of science, the historical relativity of knowledge, rules of institutional cooperation, rejection of philosophical systems, and epistemological scepticism. Fontenelle deserves much of the credit for replacing a metaphorically articulated belief in the advancement of natural philosophy with the theory of irreversible progress of the human spirit that functioned as the leading metaphor for European scientific culture from the eighteenth to the twentieth century.

Bibliography

Adorno, Theodor W. 1964. Fortschritt. In *Die Philosophie und die Frage nach dem Fortschritt*, ed. Helmut Kuhn and Franz Wiedmann, 30–48. München: Pustet.

Applebaum, Wilbur, ed. 2000. *Encyclopedia of the scientific revolution from Copernicus to Newton*. New York/London: Garland.

Barash, Jeffrey A. 1998. The sense of history: On the political implications of Karl Löwith's concept of secularization. *History and Theory* 37: 69–82.

Baron, Hans. 1959. The querelle of the ancients and the moderns as a problem for renaissance scholarship. *Journal of the History of Ideas* 20: 3–22.

Blumenberg, Hans. 1983. *Die Lesbarkeit der Welt*. Frankfurt am Main: Suhrkamp.

———. 1986. *Lebenszeit und Weltzeit*. Frankfurt am Main: Suhrkamp.

———. 1996a. *Die Legitimität der Neuzeit. Erneuerte Ausgabe*. Frankfurt am Main: Suhrkamp.

———. 1996b. *Höhlenausgänge*. Frankfurt am Main: Suhrkamp.

———. 2001. *Löwen*. Frankfurt am Main: Suhrkamp.

———. 2009. *Geistesgeschichte der Technik*. Frankfurt am Main: Suhrkamp.

———. 2013. Dogmatische und rationale Analyse von Motivationen des technischen Fortschritts. *Zeitschrift für Kulturphilosophie* 7: 407–422.

Buck, August. 1958. Aus der Vorgeschichte der Querelle des Anciens et des Modernes in Mittelalter und Renaissance. *Bibliothèque d'Humanisme et Renaissance* 20: 527–541.

Burkert, Walter. 1997. Impact and limits of the idea of progress in classical antiquity. In *The idea of progress*, ed. Arnold Burgen, Peter McLaughlin, and Jürgen Mittelstrass, 19–45. Berlin/New York: De Gruyter.

Bury, John B. 1920. *The idea of progress. An inquiry into its origin and growth*. London: Macmillan.

Crombie, Alistair C. 1975. Some attitudes to scientific progress: Ancient, medieval and early modern. *History of Science* 13: 213–220.

———. 1997. Philosophical commitments and scientific progress. In *The idea of progress*, ed. A. Burgen, P. McLaughlin, and J. Mittelstras, 47–63. Berlin/New York.

Dagen, Jean. 1977. *L'histoire de l'esprit humain dans la pensée française de Fontenelle à Condorcet*. Paris: Klincksieck.

Daston, Lorraine. 1991. Marvelous facts and miraculous evidence in early modern Europe. *Critical Inquiry* 18: 93–123.

———. 1994. Neugierde als Empfindung und Epistemologie in der frühmodernen Wissenschaft. In *Macrocosmo in microcosmo. Die Welt in der Stube: Zur Geschichte des Sammelns 1450 bis 1800*, ed. Andreas Grote, 35–59. Leske: Opladen.

———. 2001. Die Kultur der wissenschaftlichen Objektivität. In *Ansichten der Wissenschaftsgeschichte*, ed. Michael Hagner, 137–158. Frankfurt am Main: Fischer.

———. 2002. Geschichtswissenschaft und Wissenschaftsgeschichte. In *Wissenschaftsgeschichte und Geschichtswissenschaft: Aspekte einer pragmatischen Beziehung*, ed. Stefan Jordan and Peter Walther, 11–16. Spenner: Waltrop.

———. 2003a. Die Biographie der Athene oder Eine Geschichte der Rationalität. In Lorraine Daston, *Wunder, Beweise und Tatsachen. Zur Geschichte Rationalität*, 7–26. Frankfurt am Main: Fischer.

———. 2003b. Early modern history meets the history of the scientific revolution: Thought towards a rapprochement. In *Zwischen den Disziplinen? Perspektiven der Frühneuzeitforschung*, ed. Helmut Puff and Christopher Wild, 37–54. Göttingen: Wallstein.

———. 2004. Attention and the values of nature in the enlightenment. In *The moral authority of nature*, ed. Lorraine Daston and Fernando Vidal, 100–126. Chicago/London: University of Chicago Press.

———. 2011. The empire of observation, 1600–1800. In *Histories of scientific observation*, ed. Lorraine Daston and Elisabeth Lunbeck, 81–135. Chicago/London: The University of Chicago Press.

———. 2015. La philosophie naturelle. In *Histoire des sciences et des savoirs. Vol. 1: De la Renaissance aux Lumières*, ed. Stéphane Van Damme. Paris: Seuil.

Daston, Lorraine, and Peter Galison. 2007. *Objectivity*. New York: Zone Books.

Daston, Lorraine, and Katherine Park. 2001. *Wonders and the order of nature, 1150–1750*. New York: Zone Books.

Daston, Lorraine, and Katharine Park, eds. 2006. *The Cambridge history of science. Volume 3: Early modern science*. Cambridge: Cambridge University Press.

Delvaille, Jules. 1910. *Essai sur l'histoire de l'idée de progrès. Jusqu'a la fin du XVIIIe siècle*. Paris: Félix Alcan.

Dihle, Albrecht. 1969. The idea of progress in classical antiquity by Ludwig Edelstein. *Gnomon* 41: 433–439.

———. 1988. Fortschritt und goldene Urzeit. In *Kultur und Gedächtnis*, ed. Jan Assmann and Tonio Hölscher, 150–169. Frankfurt am Main: Suhrkamp.

Dodds, Eric R. 1968. Review: The idea of progress by Ludwig Edelstein. *Journal of the History of Ideas* 29: 453–457.

———. 1978. *The ancient concept of progress and other essays on Greek literature and belief*. Oxford: Clarendon Press.

Edelstein, Ludwig. 1967. *The idea of progress in classical antiquity*. Baltimore: The Johns Hopkins Press.

Feest, Uljana, and Thomas Sturm. 2011. What (good) is historical epistemology? *Erkenntnis* 75: 285–302.

Flasch, Kurt. 2017. *Hans Blumenberg. Philosoph in Deutschland. Die Jahre 1945 bis 1966*. Frankfurt am Main: Vittorio Klostermann.

Garber, Daniel, and Ayers Michael, eds. 1998. *Cambridge history of seventeenth-century philosophy*. Cambridge: Cambridge University Press.

Gembicki, Dieter, and Reichardt, Rolf. 1993. Progrès. In *Handbuch politisch-sozialer Grundbegriffe in Frankreich*, 1680–1820. Heft 14–15: *Corruption. Décadence. Ordre. Désordre. Progrès*, 101–153. Berlin/Boston: Oldenbourg.

Goldstein, Jürgen. 1998. *Nominalismus und Moderne. Zur Konstitution neuzeitlicher Subjektivität bei Hans Blumenberg und Wilhelm von Ockham*. Freiburg im Breisgau: Alber.

Gordon, Peter E. 2019. Secularization, genealogy, and the legitimacy of the modern age: Remarks on the Löwith-Blumenberg debate. *Journal of the History of Ideas* 80: 147–170.

Griffioen, Sjoerd. 2019. Secularization between faith and reason: Reinvestigating the Löwith-Blumenberg debate. *New German Critique* 46 (136): 71–101.

Groh, Dieter. 2010. *Göttliche Weltökonomie: Perspektiven der wissenschaftlichen Revolution von 15. bis zum 17. Jahrhundert*. Frankfurt am Main: Suhrkamp.

Günther, Horst. 1996. *Le temps de l'histoire. Expérience du monde et catégories temporelles en philosophie de l'histoire, de saint Augustin à Pétrarque, de Dante à Rousseau*. Paris: Éditions de la Maison des sciences de l'homme.

Hacking, Ian. 2006. *The emergence of probability. A philosophical study of early ideas about probability, induction and statistical inference*. Cambridge: Cambridge University Press.

Hadot, Pierre. 1995. *Philosophy as a way of life. Spiritual exercises from Socrates to Foucault*. Oxford: Oxford University Press.

———. 2002. *What is ancient philosophy*. Cambridge, Mass./London: Harvard University Press.

Harrison, Peter. 2019. Introduction: Narratives of secularization. In *Narratives of secularization*, ed. Peter Harrison, 1–9. London: Routledge.

Israel, Jonathan. 2004. Philosophy, history of philosophy, and l'histoire de l'esprit humain: A historiographical question and problem for philosophers. In *Teaching new histories of philosophy*, ed. J.B. Schneewind, 329–343. Princeton: Princeton University Press.

Jauss, Han R. 1964. Ästhetische Normen und geschichtliche Reflexion in der 'Querelle des Anciens et des Modernes.'. In *Parallèle des anciens et des modernes en ce qui regarde les arts et les sciences*, ed. Charles Perrault, 8–64. München: Eidos.

Kehl, Alois, and Henri-Irénée Marrou. 1978. Geschichtsphilosophie. In *Realexikon für Antike und Christentum*, ed. Theodor Klauser, vol. 10, 703–779. Stuttgart: Hiersemann.
Keller, Alex C. 1950. Zilsel, the artisans, and the idea of progress in the renaissance. *Journal of the History of Ideas* 11: 235–240.
———. 1972. Mathematical technologies and the growth of the idea of technical progress in the sixteenth century. In *Science, medicine and society in the renaissance*, ed. Allen G. Debus, 11–27. London: Science History Publications.
Koselleck, Reinhard. 1975. Fortschritt: IV: Die Ausprägung des neuzeitlichen Fortschrittsbegriffs. In *Geschichtliche Grundbegriffe. Historisches Lexikon zur politisch-sozialen Sprache in Deutschland*, ed. Otto Brunner, Werner Conze, and Reinhard Koselleck, vol. 2, 384–391. Stuttgart: Klett-Cotta.
———. 2002. *The practice of conceptual history: Timing history, spacing concepts*. Stanford: Stanford University Press.
———. 2003. *Zeitschichten. Studien zur Historik*. Frankfurt am Main: Suhrkamp.
———. 2006. *Begriffsgeschichten. Studien zur Semantik und Pragmatik der politischen und sozialen Sprache*. Frankfurt am Main: Suhrkamp.
Krohn, Wolfgang. 1976. Zur soziologischen Interpretation der neuzeitlichen Wissenschaft. In *Edgar Zilsel, Die sozialen Ursprünge der neuzeitlichen Wissenschaft*, ed. Wolfgang Krohn, 7–43. Frankfurt am Main: Suhrkamp.
Levitin, Dmitri. 2014. The experimentalist as humanist: Robert Boyle on the history of philosophy. *Annals of Science* 71: 149–182.
———. 2015. *Ancient wisdom in the age of the new science. Histories of philosophy in England, c. 1640–1700*. Cambridge: Cambridge University Press.
Lloyd, Geoffrey E.R. 1999. *Magic, reason and experience. Studies in the origins and development of Greek science*. London: Duckworth.
Loewenstein, Bedřich. 2009. *Der Fortschrittsglaube. Europäisches Geschichtsdenken zwischen Utopie und Ideologie*. Berlin: Wissenschaftsliche Buchgesellschaft.
Losee, John. 2004. *Theories of scientific progress. An introduction*. New York: Routledge.
Löwith, Karl. 1949. *Meaning in history*. Chicago/London: University of Chicago Press.
———. 1963. Das Verhängnis des Fortschritts. In *Die Idee des Fortschritts. Neun Vorträge über Wege und Grenzen des Fortschrittsglaubens*, ed. Erich Burck, 17–40. München: Beck.
Meier, Christian. 1975. Fortschritt II: 'Fortschritt' in der Antike. In *Geschichtliche Grundbegriffe. Historisches Lexikon zur politisch-sozialen Sprache in Deutschland*, ed. Otto Brunner, Werner Conze, and Reinhart Koselleck, vol. 2, 353–363. Stuttgart: Klett-Cotta.
———. 1978. Ein antikes Äquivalent des Fortschrittsgedankens: Das Können-Bewusstsein des 5. Jahrhunderts v. Chr. *Historische Zeitschrift* 226: 265–316.
Miller, David M., and Dana Jalobeanu, eds. 2022. *The Cambridge history of philosophy of the scientific revolution*. Cambridge: Cambridge University Press.
Monod, Jean-Claude. 2002. *La querelle de la sécularisation de Hegel à Blumenberg*. Paris: Vrin.
Nadler, Steven, ed. 2008. *A companion to early modern philosophy*. Malden: Blackwell.
Nasim, Omar W. 2013. Was ist historische Epistemologie? In *Nach Feierabend*, ed. Michael Hagner and Claus Hirschi, 123–144. Berlin: Diaphnes.
Neuser, Wolfgang. 2011. Fortschritt. In *Neues Handbuch der philosophischer Begriffe*, ed. Petra Kolmer, and Armin G. Wildfeuer, vol. 1, 797–801. München: Karl Alber.
Niiniluoto, Ilkka. 2019. Scientific Progress. *The Stanford Encyclopedia of Philosophy* (winter 2019 edition), ed. Edward Nouri Zalta. https://plato.stanford.edu/archives/win2019/entries/scientific-progress/.
Nippel, Wilfried. 1998. Fortschrittsgedanke. In *Der Neue Pauly: Enzyklopädie der Antike*, ed. Hubert Cancik, vol. 4, 594–598. Stuttgart: Metzler.
Nisbet, Robert. 1994. *History of the idea of progress*. New Brunswick: Routledge.
Pomian, Krzystof. 1988. *L'ordre du temps*. Paris: Gallimard.
Poovey, Mary. 1998. *A history of the modern fact. Problems of knowledge in the sciences of wealth and society*. Chicago: The Chicago University Press.

Prioreschi, Plinio. 2002. The idea of scientific progress in classical antiquity and in the middle ages. *Vesalius* 8: 34–45.

Rapp, Friedrich. 1992. *Fortschritt. Entwicklung und Sinngehalt einer philosophischen Idee*. Darmstadt: WBG.

Rheinberger, Hans-Jörg. 2008. *Historische Epistemologie zur Einführung*. Hamburg: Junius.

Ritter, Joachim. 2021. Fortschritt. In *Historisches Wörterbuch der Philosophie*, ed. Joachim Ritter, Karlfried Gründer, and Gottfried Gabriel, vol. 2, 1032–1059. Basel: Schwabe.

Rohbeck, Johannes. 1987. *Die Fortschrittsidee der Aufklärung*. Frankfurt am Main/New York: Campus.

———. 2005. Von der Heilsgeschichte zur Weltgeschiche: Histoire universelle bei Bossuet und Turgot. In *Scientia et Religio. Religionphilosophische Orientierungen*, ed. René Kaufmann and Holger Ebelt, 111–123. Dresden: Thelen.

———. 2008. Geschichtsphilosophie der Auklärung. In *Grundriss der Geschichte der Philosophie: Die Philosophie des 18. Jahrhunderts 2: Frankreich*, ed. Johannes Rohbeck and Helmut Holzhey. Basel: Schwabe.

Rossi, Paolo. 1970. *Philosophy, technology, and the arts in the early modern era*. Madison: Harper & Row.

———. 1996. Bacon's idea of science. In *Cambridge companion to Bacon*, ed. Markku Peltonen, 25–46. Cambridge: Cambridge University Press.

Rouvillois, Frédéric. 2010. *L'invention du progrès, 1680–1730*. Paris: CNRS.

Schlobach, Jochen. 1980. *Zyklentheorie und Epochenmetaphorik. Studien zur bildlichen Sprache der Geschichtsreflexion in Frankreich von der Renaissance bis zur Frühaufklärung*. München: Fink.

———. 2001. Progress. In *Encyclopedia of the Enlightenment*, ed. Michel Delon, 1100–1105. London and New York: Routledge.

Schmidt-Biggemann, Wilhelm. 1987. Säkularisierung und Theodizee – Anmerkungen zu geschichtstheologischen Interpretationen der Neuzeit in den 50er und 60erJahren. In *Auslegung des Glaubens*, ed. Ludger Honnefelder, and Matthias Lutz-Buchmann, 125–143. Berlin: Morus.

Seifert, Arno. 1982. Neuzeitbewusstsein und Fortschrittsglaube bei Leibniz. *Studia Leibnitiana, Sonderheft* 10: 172–185.

———. 1986. Von der heiligen zur philosophischen Geschichte. *Archiv für Kulturgeschichte* 68: 81–117.

Sommer, Andreas U. 2014. Sinnstiftung durch Geschichte. Die Geburt der speculativ-universalistischer Geschichtstheorien im Zeitalter der Aufklärung. In *Strukturen des Denkens. Studien zur Geschichte des Geistes*, ed. Günter Dux, and Jörn Rüsen, 203–217. Wiesbaden: Springer.

Spadafora, David. 1990. *The idea of progress in eighteenth-century Britain*. New Haven: Yale University Press.

Taguieff, Pierre. 2002. L'idée de progrès: Une approche historique et philosophique. *Les Cahiers du CEVIPOF* 32: 1–136.

Tamás, Demeter. 2015. Values, norms and ideologies in early modern inquiry: An introduction. In *Conflicting values of inquiry: Ideologies of epistemology in early modern Europe*, ed. Tamás Demeter, 1–8. Leiden: Brill.

Thraede, Klaus. 1962. Erfinder. In *Das Reallexikon für Antike und Christentum*, vol. 5, 1192–1278. Stuttgart: Hiersemann.

———. 1972. Fortschritt. In *Das Reallexikon für Antike und Christentum*, ed. Ernst Dassmann, vol. 8, 141–182. Stuttgart: Hiersemann.

Trinkaus, Charles. 1987. Antiquitas versus modernitas: An Italian humanist polemic and its renaissance. *Journal of the History of Ideas* 48: 11–21.

Van Damme, Stéphane, ed. 2015. *Histoire des sciences et des savoirs*, vol. 1: *De la Renaissance aux Lumières*. Paris: Seuil.

Wagar, Warren. 1967. Modern views of the origins of the idea of progress. *Journal of the History of Ideas* 28: 55–70.

Wagner, Peter. 2016. *Progress: A reconstruction*. Cambridge/Malden, MA: Polity.
Wallace, Robert M. 1981. Progress, secularization and modernity: The Löwith-Blumenberg debate. *New German Critique* 22: 63–79.
Wolfe, Charles, and T. and Dana Jalobeanu. 2020. *Encyclopedia of early modern philosophy and the sciences*. Cham: Springer.
Zedelmaier, Helmut. 2003. *Der Anfang der Geschichte. Studien zur Ursprungsdebatte im 18. Jahrhundert*. Hamburg: Felix Meiner.
Zhmud, Leonid. 2006. *The origin of the history of science in classical antiquity*, translated by Alexander Chernoglazov. Berlin/New York: De Gruyter.
Zill, Rüdiger. 2020. *Der absolute Leser Hans Blumenberg. Eine intellektuelle Biographie*. Frankfurt am Main: Suhrkamp.
Zilsel, Edgar. 1945. The genesis of the idea of scientific progress. *Journal of the History of Ideas* 6: 325–349.

Chapter 2
Teleology and Contingency

Contents

Abstract Many Renaissance philosophers shared an anthropocentric conception of the human position in the cosmos. This was especially true for the Renaissance Platonists, but also for other authors such as Oswald Croll and Johann Amos Comenius. One of the epistemologically striking features of Renaissance anthropocentrism was the conviction that nature or God guaranteed humans the acquisition of knowledge of the cosmos. This knowledge, however, was never an end in itself; it was always part of the journey of the human soul towards God. Therefore, in the anthropocentric concept of nature, there was no place for the idea of scientific progress. Renaissance natural philosophy thus demonstrated convincingly that the idea of the progress of knowledge was not as natural as it may seem to the modern reader. This chapter further explains the rejection of anthropocentrism by early modern natural philosophers such as Galileo Galilei, René Descartes, Pierre Gassendi, Isaac Newton, Christiaan Huygens, and Bernard de Fontenelle. The exclusion of anthropocentric teleology had important epistemological and anthropological implications. The early modern mechanism, the astronomical magnification of the universe, and the introduction of the idea of a plurality of worlds entailed a reassessment of the human position in the universe and, above all, a rejection of the idea that nature was adapted to be cognizable for humans. The reconceptualization of the concept of nature in early modern natural philosophy meant that knowledge of nature would have to be the result of long-term, gradual, slow, and collective collaboration.

D. Špelda, *The Origins of the Idea of Scientific Progress*, International Archives of the History of Ideas Archives internationales d'histoire des idées 250,
https://doi.org/10.1007/978-3-031-60526-0_2

Keywords Renaissance natural philosophy · Anthropocentrism · Early modern mechanism · Plurality of worlds · Early modern astronomy · Christiaan Huygens · Bernard de Fontenelle · Marsilio Ficino

The basic metaphysical assumption of Renaissance natural philosophy (except for atomistic currents) was anthropocentric teleology. Since this anthropocentrism took various forms, several attempts to systematize them have appeared in Renaissance scholarship.[1] In this chapter, I follow the distinction introduced by Thomas Leinkauf (Leinkauf 2009, 382–384).

In Leinkauf's view, the first form of anthropocentrism was cosmological anthropocentrism, which led metaphorically from the bottom up. According to this idea, humans were at the centre of the world. This position was not privileged cosmologically, but optically. The centre was distinguished because the observer could see the *whole* cosmos from the centre and always had a symmetrical relationship with its different parts. The point was that human beings were placed by divine providence at the centre of the world to observe, admire, and grasp the world. On the basis of this understanding, humans could express admiration for the divine work and praise the Creator. This conception naturally cohered with the idea that human being was *contemplator caeli* and the world itself was *theatrum*. I discuss this form of anthropocentrism in more detail in the third chapter.

In Leinkauf's systematization, the second form of Renaissance anthropocentrism was anthropofinalism, which led metaphorically from the top down. According to this idea, the world was created for humans. A human being was the noblest inhabitant (*incola*) of this world and the world served for human benefit (*usus*). Material nature was prepared for humans, to serve them and to be able to sustain them.[2] This concept often appeared in the context of Renaissance medicine, magic, and alchemy.

The third form of anthropocentrism was an attitude that synthesized the previous two forms, namely *ad supra* (observation of the splendour of the world) and *ad infra* (exploitation and control of the earthly world). According to this metaphysical anthropocentrism, humans were located in a privileged ontological position within the metaphysical order of the world: human beings represented the connecting element between the metaphysical levels of reality. This position was expressed in Renaissance philosophy by terms and metaphors such as *copula mundi*, *vinculum naturae*, *nodus mundi*, *epitome universi*, and *microcosmus*.

Thus, according to Leinkauf, Renaissance anthropocentric teleology can be comprehended as a combination of different elements: *contemplatio/admiratio*, *incola/usus*, and *microcosmus/macrocosmus*. Their combinations and variations constituted a commonly shared anthropological assumption in Renaissance thought that surpassed the religious denominations and became a generally accepted

[1] Kessler 1992; Daston 1995; Daston 2000; Stoffel 2001; Groh 2010, 23–130.

[2] Comenius 1964, vol. 1, 353–355; cf. Bodin 1597, 273–274.

paradigm for how philosophers understood the human condition in the world (Leinkauf 2009, 384).

Leinkauf rightly noted that what was important about human centrality was not the position at the centre of the world, but the function of that position. The location at the centre served primarily to enable humans to follow their dispositions and realize their essence (Leinkauf 2017, 134). Thus, the relationship between humans and the natural world was set up in such a way that humans might fulfil their essence with the help of nature. I address these issues because, within anthropocentric natural philosophy, time did not stand out as a relevant factor of knowledge. In the world created for human beings, there was no need to introduce time in the process of acquiring natural knowledge.

2.1 Metaphysical Anthropocentrism

In this section, I want to present two features of metaphysical anthropocentrism to demonstrate that the conditions of possibility for the emergence of the idea of prescriptive progress were lacking in this philosophical context. The first feature was that the knowledge of nature was only a part of a more comprehensive process of salvation and deification of the soul, which rose from the sensible world upwards to God and underwent an inner transformation. The second feature was that the structural isomorphy between microcosm and macrocosm precluded the emergence of anything new in the natural world, and hence ruled out the possibility of further expansion of human knowledge.

Metaphysical anthropocentrism signified that the order of nature was not contingent and that every being had its place in it. Humans had a prominent position in the order of nature because they gave it unity and coherence. Marsilio Ficino, Giovanni Pico della Mirandola, and Jean Bodin used metaphors such as *vinculum naturae*, *mundi copula*, or *nexus* to illustrate the importance of human beings in the world order.[3] Paracelsus's follower Oswald Croll summarized this view: 'Man is the tye, bond, knot, joynt, packet or bundle of all the Creatures. All things created were disposed of to him, and they respect and honor him as God's steward set over the Orchard or Garden of his world.'[4]

Paracelsus articulated the relationship between humans and the world by means of the metaphor of man as the epilogue of creation. Man was the noblest and greatest creation (*grösesst und edlest zeug*) because he was created as the last being. He was given a share in all beings because he was created from the *limus terrae*, which represented a kind of extract or leach from all the strata of creation. Man thus

[3]Ficino 2001–2006, vol. 1, 16–17; Pico della Mirandola 2019, 459; Bodin 1597, 409; cf. Blair 1997, 133; Leinkauf 2017, 1695–1710.

[4]Croll 1620a, 58, English translation in Croll 1670, 55.

contained within himself everything of the earthly and heavenly world.[5] Man was the microcosm, a 'son of the whole world' and an 'extract (*auszug*) of all creatures.' Human beings were literally made of the whole world, which nourished them as its children (Paracelsus 1924–1933, vol. I/12, 39). According to Croll, God made man the beginning and the end (*principem et finem*) of the whole world. Man was an extract from all creatures, containing the qualities of all plants and animals, and bearing in his soul the image of the wisdom of God himself. He was created last to affirm the completeness and perfection of all things. Croll concluded emphatically:[6]

> Man therefore who was the last is the most excellent and noble of all Creatures, because he hath the part of the whole world, nor is there anything in the great world which may not really be found in man. [. . .] Therefore Man the greatest miracle of Nature & most admirable Extract & kernell (*nucleus*) of the four Elements, the choycest workmanship of God and most perfect Samplar (*exemplar*) of the world, is truly every Creature, for he is all the world, and he alone hath hist honor to have resemblance with all things. . .

Another aspect of human centrality was inspired by Neoplatonism. The Renaissance Platonists and some other philosophers comprehended the natural order as a system of layers. According to Ficino, the human soul was located in a central position between the five stages of being, which it connected, and could therefore rise and fall in the hierarchy of being (Ficino 2001–2006, vol. 1, 230–236). The centrality of humans meant that they bound the different levels of being together and could move between them: descend or ascend.

The task of the human soul was not to succumb to the animal seduction of the bodily senses, but to return to its spiritual homeland and unite with God. The Platonists, therefore, emphasized that humans faced a fundamental choice in life: They had to decide between a return of the soul back to God and a fall to animal and material life (Pico della Mirandola 2019, 460–461). Ficino wrote that humans could live the life of a plant, of a beast, of man, of heroes, of demons, of angels, or the life of God: 'Every man's soul makes trial of all these in itself' (Ficino 2001–2006, vol. 4, 240–242).[7]

According to Ficino, the rational soul sought to detach itself from the body, ascend through the planetary spheres back to its origin and true goal, and immerse itself in contemplation (Ficino 2001–2006, vol. 2, 263–271). The soul's whole endeavour was to become God (Ficino 2001–2006, vol. 4, 220). This naturally meant that the process of knowing had cognitive, ontological, and moral consequences. The philosopher seeking knowledge was transformed, elevated, and deified. This claim to deification was not to be understood as a prideful desire to become God. Rather, it expressed the belief that human beings became truly human only when they regained their God-likeness.

The Platonists understood natural philosophy as part of the soul's return to God. According to Eckhard Kessler, the Renaissance Platonists believed that 'nature

[5]Paracelsus 1924–1933, vol. I/12, 36–37, cf. Sartorius von Waltershausen 1935, 35–37.

[6]Croll 1620a, 63; English translation in Croll 1670, 65.

[7]Ficino 2001–2006, vol. 4, 240–242, cf. vol. 1, 230–236.

itself, in its diversity, serves [...] not as an object of observation but rather as a stimulus for looking at the universal and for making possible a return from the diversity of the particular to the universal One' (Kessler 1994, 33). Nature, then, was not a proper object of interest. Understanding nature, that is, natural philosophy, constituted only one phase of the soul's return to God. Natural philosophy was thus one of the stages through which the soul had to pass in order to free itself from empirical reality and elevate itself to God. Pico della Mirandola explained that the knowledge of nature was only a temporary stopping point for the soul during the process of self-improvement on the way to the divine One 'marked of in a series of many steps, with the Lord seated at the top' (Pico della Mirandola 2019, 464).

The ambitions of the Platonists were always concerned with transcendence, not the earthly world. The philosopher strove to penetrate the hidden metaphysical and religious foundations of reality. The knowledge of the natural world was but a ladder that the ascended human threw behind him. As a result, the philosopher's interest in nature was to be only temporary. Ficino considered lingering too long on natural things as a contemptible *curiositas* that distracted man from divine things (Ficino 1576, 1411). Knowledge of the natural world was intended to be only a short part of the journey back to the divine One.

This understanding of natural philosophy hindered the emergence of the idea of scientific progress in several ways.

The moral conception of knowledge assumed that the degree of knowledge was proportional to moral perfection. Platonism associated knowledge with moral qualities, and therefore acquiring the truth was not morally indifferent. To attain truth, humans had to undergo somatic and spiritual transformations while ascending to God. They had to lead an ascetic way of life, to know themselves, and yet to receive God's grace (Ficino 2002, 108). Knowledge of nature thus required somatic purity and spiritual preparation. Only a morally perfect human being, to whom knowledge brought fulfilment, and happiness, could be the bearer of true knowledge (Ficino 2002, 162–163).

In such a paradigm of knowledge, there was certainly individual progress in the moral improvement and in the refinement of knowledge that ascends from material nature to ever nobler spiritual truths. However, there was no need to introduce a temporal dimension of knowledge. The Platonist philosopher was not interested in overcoming tradition, collective cooperation, patient gathering of data, or future discoveries, as the prescriptive idea of progress assumes.

Another reason for the absence of the idea of progress in the context of Renaissance Platonism resulted from the idea of the soul's ascent to God. The soul's ascent up the chain of being to the first principle was a teleological and therefore finite process. Through knowledge and purification from the material world, humans would return to the divine origin of things, and here their journey would end. This is precisely why there was no vision of further expansion of natural philosophy among the Renaissance Platonists. Ficino wrote that the goal of the intellectual movement was rest (*finis motionis intellectualis est status*). Ascending into the intelligible world, the soul was meant to view the eternal *rerum rationes*. The soul was not to be preoccupied with observing the ever-changing material world, which caused unrest in it. When the knowledge-seeking soul found the final cause of all causes, then its cognitive restlessness ceased (Ficino 1576, 676–677).

The heading of the cognizing soul thus had a clearly defined *telos*. After reaching it, the amount of knowledge did not expand. The purpose of the development of knowledge was to attain intellectual immobility: an unchanging contemplation whose justification lay in merging with beatitude. In such a context, it is hard to find conditions of possibility for the idea of gradual, methodical, and indefinite progress in the knowing of nature.

The third reason for the absence of the idea of progress among the Renaissance Platonists was connected with the belief in the human vocation to understanding nature. According to the Renaissance Platonists, humans by nature desired knowledge (including knowledge of nature) and the cosmos was set up in such a way that virtuous humans could fulfil this desire. Ficino explicitly argued that humans were given a natural desire for knowledge (*naturale desiderium*) that would not cease until reaching its goal, which was knowledge of God. Unification with God was the ultimate goal of human existence and also the definitive end of the natural desire for knowledge.[8]

The success in the fulfilment of this desire was guaranteed by the overall metaphysical framework of (Neo)Platonic metaphysics. Indeed, Ficino added that one of the signs of a well-ordered creation was that all kinds of beings and entities were capable of achieving their appointed goal (Ficino 2001–2006, vol. 4, 232). Ficino therefore could say that the desire for knowledge was not given to man in vain, but in order to be fulfilled (Ficino 2001–2006, vol. 4, 238). This desire also concerned nature. According to Ficino, God wanted himself to be known through nature, which gradually awakened the innate concepts (*rationes*) in the fallen soul (Ficino 1576, 1350).

Nature thus served as a source of incentives actualizing the cognitive potential that God gave to humans. Although the Platonist did not seek to know empirical nature per se, he believed that there was a preset teleology between human cognitive faculties, empirical nature, and the realm of transcendence. In such a complex set of ideas about the interconnectedness of nature, the human soul, and God, it was difficult to think about the progress in the human knowledge of nature. The desire for knowledge of nature was quickly fulfilled and had the status of a passing phase in the realization of the higher task of the salvation and deification of the soul.

If knowledge of nature was conditioned by a moral purity that was judged and rewarded by God, knowledge became a gift – and that is another reason for inconceivability of the idea of progress. According to the Platonists, God wanted to be known, and therefore planted the seeds of wisdom in human souls. The seeds germinated after the souls received appropriate stimuli, some of which originated in nature (Ficino 1576, 1350). If such wisdom were not given to us by God, we could not discover it, wrote Croll (Croll 1620a, 63). God directly gave knowledge to humans or showed them grace by enlivening their hidden dispositions towards

[8] Ficino 2001–2006, vol. 4, 226–228; similarly Plethon 1858, 40–42.

wisdom. The idea that knowledge (or the actualization of a disposition to wisdom) was a gift resulting from God's grace hindered the conceivability of the idea of scientific progress in two ways. The idea of a gift implied closedness and finality. Knowledge that was a gift from God could hardly be further perfected. Furthermore, the idea of scientific progress presupposes that humans develop knowledge autonomously. As limited and fallen beings, they proceed slowly and by their own methods that provide neither the speed nor the extent of divine knowledge. Acquired knowledge, however, was completely the result of human effort and inventiveness.

The final reason for the absence of progress in natural philosophy inspired by Platonism was that the subject of knowledge was always the individual who sought salvation and happiness. In the Platonist context, there was no place for a long-term collective collaboration that transcended the individual life nor for a collective subject such as an institution developing knowledge in history. The philosopher knew the world for himself and the sake of his soul. The natural world responded to individual desires for knowledge and requests for help: Nature knew that individuals demanded something of it.

Let us now turn to another conceptual mean that the Renaissance philosophers used to articulate the central role of humans in the order of nature: the doctrine of microcosm and macrocosm. The human being as a microcosm encompassed the whole cosmos in both somatic and spiritual senses. Johannes Comenius wrote that 'man represents the whole world with his body and the Creator of the world with his soul. That is why he is not inappropriately called *microcosmus* and *microtheos*. *Microcosmus*, because: 1. Whatever the world has, it has for man's sake, to serve him [...]. And everything renders him service. The heavens revolve for him, the stars shine for him (*caelum illi se volvit, sidera illi lucent*)'.[9] According to Croll, man was the most significant and sublime creation because he contained all the parts of the world within himself and there was nothing in the macrocosm that could not be found in humans. Therefore man was also the greatest miracle of nature.[10]

Renaissance philosophers presumed an inner interdependence between the content and structure of human souls and bodies on the one hand, and between nature and the cosmos on the other. There was a structural isomorphy between humans and cosmos (Bianchi 1994, 29). Assuming this teleological relationship, philosophers could insist that without knowing themselves humans could not understand the world. Croll asserted that humans came to know the world and everything in it by turning inwards to their inner selves, which mirrored the macrocosm (Croll 1620a, 53). Similar statements about the relationship between microcosm and macrocosm

[9]Comenius 1964, vol. 1, 376a, similarly Khunrath 1609, 294.

[10]Croll 1620a, 63: *Ergo Homo ultima Creatura praestantissima & nobilissima est, quia totius mundi partes habet, & nihil est in maiori mundo, quod in ipso Homine realiter non reperiatur.* [...] *Homo itaque miraculum naturae maximum.*

can be found in other Renaissance writers, such as Ficino, Cornelius Agrippa, Paracelsus, Bodin, Giambattista della Porta, and Robert Fludd.[11]

An important consequence of the structural isomorphism between microcosm and macrocosm was the impossibility of the emergence of something new. The two entities mirroring each other, the microcosm and the macrocosm, were created complete and finished. If something were to suddenly emerge in the microcosm or the macrocosm, it would have to involve a transformation of the other entity due to the isomorphism of the two. Croll insisted that human anatomy replicated the anatomy of the world: indeed, they were the same.[12] That meant for instance that planets corresponded to the parts of human body (Ficino 2002, 274–275). Therefore the discovery of a new planet should lead to the discovery of a new organ in the human body that would be associated with that planet (and vice versa). Neither Renaissance metaphysics nor anatomy, however, assumed that anything hitherto unknown (such as valves and spermatozoa) would be discovered in the human body. Similarly, astronomy assumed that the sky encompassed the totality of the visible and that humans could not expect to discover new planets or stars within the natural order (comets or new stars were considered supernatural phenomena). In a world where the macrocosm completely mirrored the microcosm, the conditions for further advancement and expansion of knowledge were lacking.

This feature of Renaissance thinking about nature was related to the fact that for most Renaissance thinkers, nature was an ordered whole of parts that were interconnected. The world was not a mere sum of independent parts, but a closed whole, the cosmos whose parts were interconnected by some metaphysical principle, the nature of which varied from author to author (*amor*, *vinculum*, *spiritus*, *catena*, *sympatheia*, *connexio*, *cohaerentia*, *consensus*).

For instance, Ficino wrote in *De vita* that the world was a perfectly interconnected and living unified whole: ‘the most perfect animal’ (Ficino 2002, 250–251). As Leinkauf summarized, when the world is understood as a coherent whole of interconnected parts, nothing can be added to it, nor can anything be subtracted from it, without disrupting this order. Each part has meaning and purpose in the order of the world.[13]

In his *Theologia Platonica* (1482), Ficino questioned Lucretius’s atomistic conception of nature and emphasized the universal teleology of the natural order where all things were interconnected. That was why the parts of the order could not be added or subtracted at will: ‘. . .all the world’s parts come together to form for the whole world a unique harmonious beauty from which nothing can be subtracted and to which nothing can be added’ (Ficino 2001–2006, vol. 1, 202–203). The parts of the world might not be perfect in themselves, Ficino continued, but the arrangement

[11] Ficino 2002, 274–276; Agrippa of Nettesheim 1533, 28, 284; Paracelsus 1924–1933, vol. I/1, 203; Bodin 1597, 410–411; Porta 1597, 12; cf. Schmidt-Biggemann 2005.

[12] Croll 1620b, 25–26; cf. Hirai 2016, 382.

[13] Leinkauf 2017, 1695–1696. Leinkauf cites as an example another passage from Ficino’s *Platonic Theology*, see Ficino 2001–2006, vol. 1, 206–207.

of the whole was perfect: the world could not be better ordered (*non aliter mundum posse bene disponi quam ita*) (Ficino 2001–2006, vol. 1, 206–207).

Julius Caesar Scaliger put forward a similar argument. In his *Exotericarum Exercitationes* (1557), he asked whether everything was created for man (Scaliger 1558, 321–325). As the whole book was a critical examination of Girolamo Cardano's treatise *De Subtilitate* (1551), the question was a reaction to a corresponding passage in Cardano's work. Scaliger agreed with Cardano that the heavens revolved for man's sake (*propter homo caelum actiones*). However, Scaliger was much more interested in another of Cardano's questions, namely, whether all kinds of beings were useful to man (Cardano 1663, vol. 3, 549a).

Scaliger's standpoint was well summarized by Kuni Sakamoto: Humans were in a central position in the chain of being, between the heavens and the sublunary world. All beings, by forming a chain, helped to maintain the unique position of humans who mediated between the two worlds. In this sense, all species helped to preserve the ontological structure of the world. Sakamoto explained that in Scaliger's view not all species served humans directly. Some functioned primarily to prevent an empty place in the chain (Sakamoto 2016, 61). No being was therefore purposeless, each had a purpose and served humans directly, or indirectly.

The cosmos was an ordered unified whole in which everything had an irreplaceable ontological function. Therefore, Scaliger disagreed with Cardano's view that a fly was an insubstantial ornament to the order of the world. I quote Scaliger's explanation from Sakamoto's translation:[14]

> [The fly was made] not for the ornamentation, but for the completeness [of the world]. For the series of beings is so great and the affinity [between them] is so close that if its smallest part were absent or to be missing from [the chain of] causes, the world would cease to be a world. [. . .] If one thing were absent, everything would slide and collapse.

In Scaliger's view, if even such a miserable being as a fly were missing, the order of nature would collapse. One can assume that the same reasoning would apply if something new were to appear – a new planet, a new organ, or the invisible *animalcula*. In Scaliger's conception of the natural order, each individual had a function, a role, and a meaning in the world order. Nothing was contingent. All things formed an ordered whole. When individual parts were interrelated and interpenetrating (*omnia in omnibus*) there was no space for the emergence of new natural objects (if everything remained within the natural order).

Ficino's and Scaliger's considerations show that at least some Renaissance philosophers did not envisage the possibility of discovering radically new objects. On the contrary, the idea of expanding the world with unexpected entities implied a contingency that threatened the balanced order that was a fundamental principle of the Renaissance conception of nature.

[14] Scaliger 1558, 325; Sakamoto 2016, 64.

2.2 Anthropofinalism

The main thesis of anthropofinalism was simply summarized by the Renaissance poet Pontus de Tyard: 'everything in the world was made for the use and service of man' (Tyard 1557, 6). Such an idea naturally correlated with the idea of the cosmos being made of interconnected parts. The focal point in which all relationships intersected was a human being. At least in the context of philosophies with Platonist traits, humans had the ability to become everything and could move on different layers of a hierarchically conceived natural order. Precisely because humans were central to both the natural and metaphysical order, they did not need to gain power over nature. Following the traditional hermetic notion of man as a second God (Rüfner 1955), Ficino enthusiastically described man's dominion over the natural world:[15]

> [M]an uses all the world's materials and uses them everywhere as though they were all subject to him: he uses the elements, stones, metals, plants, and animals, fashioning them into many forms and figures, which the beasts never do. Nor is he content with just one of the elements or with particular elements like the brutes: he employs them all as though he were the lord of all. [...] In inhabiting all the elements and cultivating them all, he performs the office of God.

Dominion over the earthly world did not satisfy the desire of humans, who wanted to subdue the spiritual world: humans strove to attain the divine state (Ficino 2001–2006, vol. 4, 250). In his *Amphitheatrum sapientiae aeternae* (1595), Heinrich Khunrath also believed that 'truly immense and more than admirable is the power of man endowed with reason, who rules over all creation by the will and command of God' (Khunrath 1609, 42). Renaissance writers did not yet think that natural philosophy made humans rulers of the world. Humans obtained power over nature through God, and that was why their ambitions did not concern nature. For that reason, Renaissance natural philosophy did not consist of extending and accumulating empirical findings as means of conquering nature. And that is also why the idea of progress was lacking in this context.

Anthropofinalism was also apparent in specific epistemological and anthropological issues. A good example is the theory of signatures. Renaissance natural philosophers assumed the existence of invisible relations between things that formed a complex network of similarities, sympathies, and antipathies. The doctrine of *signatura rerum* explained the question of how to know these occult relations of things. According to the doctrine, the invisible essence of a thing, its power, or its ability to influence other things, was unveiled in the outward appearance of the thing. Every natural object revealed its hidden properties and abilities through its visible surface. The outside was the sign of the inside. Knowledge of the properties of a thing, therefore, consisted of discerning the sign that proclaimed the inner and

[15] Ficino 2001–2006, vol. 4, 172–174; cf. McKnight 1991, 42–50.

hidden properties.[16] According to Paracelsus and Croll, God did not want his gifts to remain hidden; he marked things with visible signs.[17]

Famously, Paracelsus considered the external shape of plants or their parts as signs revealing their hidden healing effects. Therefore, 'according to his theory, nature is constituted as a realm of absolute transparency, in which there is nothing so concealed that it may not be revealed by its external signs and be known by them' (Bianchi 2004, 200). Massimo L. Bianchi drew attention, for example, to a remark in Paracelsus's *Astronomia magna* (1537), in which the Swiss physician wrote: 'No mountain, no rock is so dense that it may hide and conceal what is in it, so that is not manifest (*offenbar*) to man. This, in fact, derives from its *signatum signum*.'[18]

In *De natura rerum* (1572), which was probably not authored by Paracelsus, it is also stated that all of God's creation served to benefit humans. God provided every thing with visible signs unless he judged that it should remain completely hidden from humans.[19] It is not surprising that in his *De signatura rerum* (1609), Croll also emphasized that nature itself showed man the occult properties of things, especially the healing powers of plants: 'For every thing that is intrinsic, bares the external figure of its occult property, as well in insensible as sensible Creatures. Nature as it were by certain silent notes speaks to us...'[20]

The notion of signs offered to humans naturally had significant epistemological implications. The search for an understanding of nature did not require any particular method except for hermeneutics. Nature constituted a comprehensive system of analogies, references, and signs that referred to each other, the outer qualities reflecting the inner, and the earthly qualities expressing the heavenly. The visible things referred to the invisible or formed a connected chain that served both as a way of knowing the levels of being and as a way to the Creator (Croll 1620b, 36–37). Recognition of these analogies and references was a symbolic return to the knowledge of ancient sages or to Adam's knowledge and his dominion over nature, which was based on knowledge of the true names of things (Croll 1620b, 38–39). The acquisition of knowledge concerning the forces of nature was therefore in fact a return to the original harmony between thought, being, and language.

Signatura rerum theory demonstrated that the temporal regime of knowledge was far from obvious. Nature was finite and closed, therefore the number of signatures was also closed, and therefore there was no potential for further advancement of knowledge. Also, knowledge of signatures was not based on long-term empirical investigation. The signatures were created by God and made available to humans partly as visible appearances and partly in ancient books. Nature spoke to the virtuous lovers of knowledge and opened itself up as a realm of signs accessible to all initiates. To acquire knowledge, the philosopher needed virtue, piety, God's

[16] Useful summary of the theory was provided by Hirai and Kikuchihara 2015.

[17] Paracelsus 1924–1933, vol. I/11: 393; Croll 1620b, 28.

[18] Paracelsus 1924–1933, vol. I/12: 174–175. I quote Bianchi's translation, see Bianchi 2004, 201.

[19] Paracelsus 1924–1933, vol. I/11, 393; cf. Hirai 2019.

[20] Croll 1620b, 18–19; English translation in Croll 1670, 34.

grace, and ancient books rather than patient data collection accompanied by collective discovering of the new phenomena.

Another aspect of anthropofinalism relevant to our inquiry was the assumption that nature knew about humans. According to Croll, nature had perfectly provided mankind with medicines, even going so far as to create appropriate remedies in every region, for every nation, and every epoch. Therefore, Croll discouraged bringing exotic medicines originating in distant lands, as they were primarily intended for the local population. The gracious God had ordained that at home everyone should find plants that would satisfy their needs (Croll 1620b, 19–20). Nature was therefore adapted to benefit the people who live in it and to provide remedies for their diseases.

In the view of Renaissance Paracelsians, nature was aware of more than the human race or individual nations. Nature also knew about each individual and their needs. This of course applied to all forms of astrology, which assumed a relationship between the individual and the stars (Zambelli 2007, 115–182). There was also a well-known Paracelsian recipe for a healing salve that was supposed to heal wounds sustained in battle. This salve was not to be applied to the wound itself, but to the weapon that caused it (Paracelsus 1924–1933, vol. I/14, 448). The belief in the functionality of the salve was consistent with the belief that humans were embedded in a vast web of sympathetic relationships linking macrocosm and microcosm. Nature was an organism, a universal harmony of the world that was aware of all its parts. Things in the world were interconnected by relationships that made it possible to soothe the pain of specific individuals stemming from specific wounds inflicted by specific weapons. The world of Renaissance Paracelsians thus knew about the suffering of individual human beings. It was not the universe of Cartesian and Newtonian natural philosophy, which was indifferent to humans, and in which stab wounds were merely the physical interaction of material corpuscles.[21]

In sum, for the Renaissance Platonists and many other Renaissance thinkers, knowledge of nature did not require time. The person who was the *imago Dei*, who wanted to ascend to God and who was obeyed by the natural world, did not need time to come to knowledge of nature. In this context, the truth was the result of *contemplatio Dei*, divine inspiration, affiliation with an esoteric tradition, intellectual concentration, and somatic renunciation. This philosophical paradigm also did not assume that humans had to wait patiently for knowledge or that they had to participate collectively in research and emancipate themselves from prejudice. Time began to act as a condition of knowledge only at the moment when the disparity between the cognitive capacities of humans and the immensity of nature became apparent.

Another example of the anthropofinalist standpoint including its epistemological consequences can be found in Renaissance astronomy. Like many of their contemporaries, Renaissance astronomers believed that the world was created for humans.

[21] Descartes 1996, vol. 8, 321. In his *Scepsis scientifica*, Joseph Glanvill stated that Kenelm Digby attempted to explain the functioning of the ointment on the basis of the corpuscular theory of matter; see Glanvill 1885, 178.

For astronomers, this did not only mean that humans were the centre of the order of nature nor that nature spoke to them through signs. In astronomy, anthropocentrism meant the accessibility of the mathematical order of the cosmos to human thought. The world created for the sake of humans was intelligible because it was a geometrical order.

In the introduction to his *De revolutionibus orbium coelestium* (1543), Copernicus wrote that the world was created for man and according to human standards of geometric harmony. Explaining the origins of his cosmology, Copernicus reproached the philosophical and astronomical tradition for producing constructions that did not correspond to the geometrical order of the universe: 'I began to be annoyed that the movements of the world machine, created for our sake (*propter nos*) by the best and most systematic Artisan of all, were not understood by the philosophers…'[22]

The *propter nos* formula, as Blumenberg observed, referred to the natural intelligibility of the world. The anthropocentric teleology of the Copernican cosmos was manifested precisely in the fact that it was open to human desire of knowledge and provided a path to happiness and knowledge of God. The world was made for human beings in the sense that human cognition could penetrate the sensory illusion of geocentrism and understand the true arrangement of the heavenly bodies (Blumenberg 1996a, 50–52).

Tycho Brahe also believed in the divine symmetry of the cosmos, where the parts of the world were proportionate to each other, including the distances and sizes of the heavenly bodies. The cosmos was ordered with respect to humans and their cognitive abilities. Human ideas of symmetry and proportion could thus be a reliable measure of the beauty and rationality of the cosmos.[23]

Johannes Kepler also based his interpretation of the anthropocentric teleology of creation on the idea of the intelligibility of the world. As early as in his work *Mysterium cosmographicum* (1596), Kepler insisted that man was the purpose of creation. When creating the world, God had in mind its future inhabitant.[24] Everything in the world was created for the sake of humans (*omnia propter hominem*) (Kepler 1938–2009, vol. 1, 79). Kepler retained this belief for the rest of his life. In his late treatise *Epitome astronomiae copernicanae* (1618–1621), he wrote: 'For the Earth was going to be the home of the speculative creature, and for his sake, the universe was made' (Kepler 1938–2009, vol. 7, 277).

Kepler was influenced by Neoplatonism, Hermeticism, and Pythagoreanism and he used an exemplarist metaphysics to justify the epistemological accessibility of the world. The cosmos was created perfectly according to sample ideas that Kepler called archetypes and that he believed belonged to the divine essence (Kepler 1938–2009, vol. 6, 360). These ideas were geometrical in nature, were coeternal

---

[22] Copernicus 1975, 4; English translation in Copernicus 1978, 4; cf. Lerner 2008, vol. 2, 69–73.

[23] Brahe 1913–1929, vol. 6, 222, cf. vol. 2, 435, vol. 4, 155–157; see also Granada 2013, 195–196.

[24] Kepler 1938–2009, vol. 1, 30: *…nemo negabit, in domicilio mundi exornando Deum ad incolam futurum identidem respexisse. Finis enim et mundi omnis creationis Homo est.*

with the Creator (*geometriae rationes Deo coaeternae sunt*), and formed the basis for the creation of the world.[25] When Kepler's God created the world, he practiced geometry.[26] The harmony, beauty, and rationality of the world consisted primarily of embodied geometry. Human beings were meant to transcend the sensory image of the cosmos and rationally discover its mathematical foundations. In his *Epitome*, Kepler presented astronomy 'as a discipline which discloses the causes of things, shakes off the deception of eyesight and carries the mind higher and farther, outside the boundaries of eyesight.'[27]

While the world was created according to mathematical archetypes, the human mind was created to be able to understand this geometric structure:[28]

> God, who founded everything in the world according to quantitative norms, has endowed man with a spirit that can understand them [...] The spirit of man is created not for the knowledge of arbitrary things, but for the knowledge of quantities; he understands a thing the more correctly the more that thing approaches the pure quantities that are its origin.

For Kepler, the metaphysical guarantee for the human mind's ability to understand the structure of reality was primarily the theological doctrine according to which humans were created in the image of God, a view he shared with other Protestant thinkers of the period. 'The world is the bodily image of God; the soul is the disembodied, created image of God.'[29] In Kepler's view, humans resembled God in their ability to grasp the basic (mathematical) truths of creation. God gave his human images the disposition to understand the laws and numerical and geometrical relationships that constituted the divine mathematical plan of the world: 'God willed that we should know them [i.e., geometrical notions] when he made us in his image, and thus we received a share in his own thoughts.'[30]

According to Kepler, our quantitative knowledge, 'as far as piety permits us to say, is of the same kind as God's.'[31] If humans knew the world by means of numbers and geometry, they knew the world in the same way as God, for their competency in geometrical thinking was the same. When thinking mathematically and rationally, humans – as images of God –participated directly in divine thought. Naturally, the human mind could never grasp the absolute truth in all its divine perfection, for there were areas that were inaccessible to human knowledge. Kepler noted that humans

---

[25] Kepler 1938–2009, vol. 7, 267, cf. vol. 6, 104; cf. Mehl 2013.

[26] Kepler 1938–2009, vol. 6, 299: *Non aberat* [...] *ab archetypo suo Creator, Geometriae fons ipssimus, et, ut Plato scripsit, aeternam exercens geometriam.* Cf. Barker and Goldstein 2001; Methuen 1998.

[27] Kepler 1938–2009, vol. 7, 258; English translation in Kepler 1995, 14.

[28] Kepler, letter to Mästlin, 19 April 1597, Kepler 1938–2009, vol. 13, 113: *mens hominis non ad quaevis, sed ad quanta intelligenda condita est, remque quamlibet tanto rectius percipit, quanto illa propior est nudis quantitatibus, ceu suae origini.*

[29] Kepler, letter to Hervart, 9 April 1599, Kepler 1938–2009, vol. 13, 312; cf. Groh 2003, 592–594, 607–609, 712–714, 738–742.

[30] Kepler, letter to Hervart, 9 April 1599, Kepler 1938–2009, vol. 13, 309; cf. Cassirer 1922, vol. 1, 328–370; cf. Barker 1997.

[31] Kepler, letter to Herwart, 9 April 1599, Kepler 1938–2009, vol. 13, 309.

had only a partial share in divine thoughts, but this share concerned precisely the geometrical structure of the world.[32] The origin of such knowledge, however, was not empirical but *a priori*, for the basic mathematical truths were innate to the human mind and their 'awakening' occurred after investigating the heavenly phenomena (Kepler 1938–2009, vol. 6, 228).

This epistemology resulted in Kepler's attitude towards the role of time in the process of knowledge. The truth of the cosmos was timeless: the eternal mathematical archetypes embodied in creation by God were the unchanging foundation of the cosmos. The human mind could grasp them *a priori* because it was prepared for this task. The mind needed empirical data as a stimulus for geometrical reasoning rather than as data ready for inductive generalization.

We can see this Neoplatonic mentality in *Dissertatio cum Sidereo Nuncio* (1612), where Kepler responded to the greatest discovery of the time – the telescopic observations made by Galileo with his telescope. While Kepler appreciated the epochal significance of Galileo's discoveries, he also immediately relativized their importance in light of his exemplarist metaphysics and rationalist epistemology. According to Kepler, those who know the causes of things before they are observable by the senses are certainly closer to the Creator – unlike those who only think about causes on the basis of experience. Platonists and Pythagoreans thus knew the true order of the cosmos long before the beginnings of empirical astronomy which means that they correctly presumed what Galileo had discovered by observation (Kepler 1938–2009, vol. 4, 305). With the rigour of a metaphysician, Kepler insisted that the affinity of the human mind with the divine mind in the mathematical *rationes* resulted in its independence from experience, which played only the secondary role of a stimulus in the process of acquiring knowledge. Galileo was a mere empiricist and mechanic, while Kepler presented himself as a *philosophus* who recognized causes because he was closer to God thanks to his mind.[33]

In Kepler's Platonic paradigm, there was no space for the idea of scientific progress. Astronomy naturally cannot proceed without long-term observations that provide the data necessary for calculations. The real purpose of astronomy, however, was to understand the causes of celestial motions and the geometrical structure of the cosmos. The human mind accomplished this task by rational reasoning, not by collecting observations over a long period of time. Similar to God, humans shared some of his thoughts, and the world opened up to them in its geometrical arrangement, so there was no need to wait for the truth. From a purely epistemological point of view, truth was not a function of time. The truth of the cosmos was eternal and unchanging. There was no space for continual gradual and cumulative deciphering of the secrets of cosmos. Humans either grasped the geometrical structure of the cosmos as a whole or they did not understand it at all. This truth has come and gone throughout the history of humankind, but its occurrence depended on the grace of God, not on a cumulative and long-term process of knowledge. Similar to other

[32] Ibid., 312; cf. Kepler 1938–2009, vol. 7, 293.

[33] See Bucciantini 2008, 230–242; Hamou 1999, vol. 1, 183–190.

Protestant astronomers (Groh 2010, 465–550), Kepler believed that the revelation of truth was part of the historical plan of Providence. The idea of scientific progress was hardly conceivable in this epistemological context because it implied the human authorship of knowledge.

2.3 Rejection of Anthropocentrism: Metaphysical Contingency

The metaphysical opposite of teleology was contingency. Three forms of contingency were the most important to early modern natural philosophy: metaphysical, cosmological, and anthropological contingency.

The main source of the idea of metaphysical contingency was mechanism. In recent scholarship, the early modern mechanism has been intensively investigated and historians have corrected many older and simplistic interpretations. In particular, it has been shown that there were several forms of mechanisms in the early modern period. The various currents agreed on the need to reject Aristotle's natural philosophy but differed in their methods, assumptions, and intentions.[34] Historians today mostly agree that mechanistic ontology should be separated from the explanatory (or methodological) mechanism that sought mechanistic explanations for specific phenomena without being strictly committed to mechanistic ontology.[35]

Stephen Gaukroger devoted two volumes to the context, emergence, spread, and collapse of mechanism[36] and I take advantage of his characterization of ontological mechanism here. Gaukroger pointed out that mechanism has existed in many varieties, yet an 'ideal-type mechanism' can be constructed that 'reduces all physical processes to the activity of inert corpuscles making up macroscopic objects, where the behaviour of these corpuscles can be described exhaustively in terms of mechanics and geometry, and where they act exclusively by means of efficient causes, which require spatial and temporal contact between the cause and the effect.'[37]

The ontological mechanism mostly, but not exclusively, assumed that particles of matter were passive. The question of how the invisible and inert material particles made up the visible complexity of nature was a matter of debate in seventeenth-century natural philosophy. The mechanists firmly rejected the notion that any active principles or occult qualities, as found in Aristotelian philosophy or Renaissance philosophy, were the source of the motion and order of corpuscles. They mostly supposed that God was the source of activity and the power shaping the order of nature. At this point, then, the ontological mechanism merged with theological

---

[34] See Roux 2017, 30–33; Hattab 2019; Roux 2022, 75–80.

[35] Des Chene 2005, 246; Des Chene 2000b, 14; cf. Roux 1998; Roux 2017; Meli 2019, 3–7, 23–24; Gaukroger 2010, 6–7; Walsh 2019, 149; see also Clarke 1989, 190; Clarke 2018, 101–110.

[36] Gaukroger 2006 and Gaukroger 2010.

[37] Gaukroger 2006, 260, cf. also definitions in Gaukroger 2010, 58, 98.

voluntarism and presented itself as authentic Christian philosophy, in contrast to the scholasticism and Renaissance philosophy that were, according to mechanists, infected with the spirit of pagan doctrines.

The explanatory mechanism was primarily a methodological programme. According to Steven Nadler, the explanatory mechanism sought to explain the properties and behaviour of an object by means of its internal structure, that is, the particles that constitute the object. Mechanists regarded the internal corpuscular structure of objects as the cause of the perceived properties and behaviour of the object. Therefore they also considered mechanistic explanations as causal. Further, as the assumed corpuscular structure of objects was imperceptible, it was not amenable to empirical confirmation, and therefore it was regarded as only hypothetical (Nadler 2000, 531–536). Finally, it was typical of the explanatory mechanism that it refused to explain the behaviour of things in terms of final causes and demanded only an explanation based on efficient causality (Osler 1994; Daston 1995).

The point of distinction between ontological and explanatory mechanisms was put succinctly by Dennis Des Chene: 'Mechanism as ontology failed, but mechanism as method succeeded.'[38] The ontological mechanism served as an alternative ontology to Aristotle's philosophy and contributed significantly to its rejection. It failed, however, to explain chemical processes, electricity, and above all, the processes of the living world: from the birth of living beings to the action of muscles. The explanatory mechanism proved to be an effective means of making nature intelligible. This programme removed Aristotelian metaphysical categories and occult qualities from natural philosophy and replaced them with comprehensible micromechanical models of visible phenomena. Mechanicists, however, remained sceptical about the possibility of revealing the causes and essences of things.

In the second half of the seventeenth century, therefore, more and more representatives of the new non-Aristotelian natural philosophy resorted to mechanistic explanations that did not strictly adhere to the mechanistic corpuscular ontology. For example, various active or vital principles began to be (re)introduced into explanations of the manifestations of life. In physics, Newton's notion of gravitational force was controversial because it was not compatible with the basic tenets of mechanistic theory of matter.

In the following paragraphs, I want to show some features of the ontological mechanism that were not compatible with an anthropocentric teleology of nature. Based on this explanation, I further strive to demonstrate that the mechanistic reconceptualization of nature opened way to conditions of possibility for the idea of scientific progress.

Mechanists rejected the idea that nature was capable of existing, causally acting, and persisting in itself. In British natural philosophy, terms such as 'inanimate brute matter' (Newton 1961, 249, 253), or 'senseless and stupid matter' (Boyle 1775, vol.

[38] Des Chene 2005, 249. Gaukroger disscussed this process thoroughly in his books: see Gaukroger 2010, 293–327, Gaukroger 2016, 70–120.

5, 511) were used. But a thorough metaphysical justification of the causal impotence of nature had already been provided by Descartes (Descartes 1996, vol. 8, 61).

In the third meditation, Descartes famously introduced his idea that God preserved (*conservet*) things in their existence (Descartes 1996, vol. 7, 369). Descartes denied that the *res extensa* carried any intrinsic meaning that ensured its continued existence in the sense of moving towards the fulfilment of its purpose. Created objects were incapable of existing on their own over time. As Blumenberg succinctly explained, for everything conditioned by a creative act in its first moment of existence, the lack of intrinsic necessity applied in subsequent moments (Blumenberg 1970, 5). Created substances were indifferent to their own existence and that is why their continued existence had to be provided by God in the form of *conservatio*.[39]

Unlike the cosmos of Renaissance philosophy, Cartesian reality bore no *vis per se existendi* and was fully dependent on the power and will of a transcendent Creator. For Renaissance philosophers, the world did not disintegrate into random and discontinuous states, because it was a stable hierarchical order of catenated beings. The continual existence and coherence of the Cartesian world was ensured by *concursus ordinarius*: The immutable God preserved the laws that he imposed on his creation at the beginning, but nature itself was endowed neither with order nor with the capacity for self-organization.[40]

The idea of contingent nature formed an important part of the mechanistic natural philosophy in the second half of the seventeenth century. Boyle, for example, asserted that 'no being but God can subsist by its own power, but must be continually upheld by the constant influx of the conserving power of God.'[41] Samuel Clarke made a similar point in his polemic with Gottfried Wilhelm Leibniz (Leibniz and Clarke 1978, 353–354, 361) and ultimately so did Newton himself.[42]

I do not want to deal here with the different variants of mechanistic ontology, but mainly with its epistemological and anthropological consequences. If nature were contingent and dependent on God's power, the human mind would lose the guarantee of its knowability that Renaissance philosophy presupposed. The Renaissance notion of an animated and rational cosmos that was adapted to human cognitive capacities was rejected by representatives of mechanistic ontology as a manifestation of pagan necessitarianism. The contingent world was not made accessible to human minds, and humans were not adapted to understand the order of nature.

In the mechanistic paradigm, the belief in the close relationship between human mind and the order of nature established by God (according to eternal ideas) vanished. The Renaissance *propter nos* formula became not only a prejudice but also an impiety that restricted the freedom of the divine will to human interests and

---

[39] Descartes, letter to Hyperaspitas, August 1641; Descartes 1996, vol. 3, 429.

[40] Descartes 1996, vol. 8, 62, cf. especially Gaukroger 2010, 106–108; Garber 1992, 266–287, Garber 2001, Schmaltz 2008, 87–89, 102–104; Des Chene 2000a, 325–341.

[41] Boyle 1775, vol. 6, 697, vol. 5, 163, 176–179; vol. 2, 403; vol. 5, 68; cf. Shanahan 1988.

[42] Newton 1726, 528; Newton 2004, 138.

needs. According to Descartes, God created the immense world primarily for his own sake and for the sake of realizing his own purposes: *omnia propter ipsum facta sunt*. God was both the final and efficient cause of the universe, and no being could claim that it was for her that creation was intended.[43] The reasons for the decisions of divine will were inaccessible to man. When creating the world, God did not have to obey any rules that limited his free will and absolute power.[44]

Blumenberg noted that in the moment when philosophy emphasized the absolute power of God, the natural world became 'a pure performance of reified omnipotence', 'a demonstration of the unlimited sovereignty of a will to which no questions can be addressed' and this meant 'that, at least for man, the world no longer possessed an accessible order'.[45] Mechanistic natural philosophy thus grasped the creation of the world in the same way as theological voluntarism,[46] according to which the natural world was a fact of reified omnipotence, to borrow Blumenberg's expression. There was only one answer to the questions concerning God's creative act formulated as early as by Augustine: *quia voluit*.[47]

The incongruity between human thought and the will and omnipotence of the Creator was the crucial epistemological element of early modern natural philosophy; therefore, it was often mentioned in several identifiable variants.

In epistemology, the voluntarist position meant that the limits of human thought must not in any way restrict divine power. The world was in proportion to the Creator's power, not to human thought. This notion already occurred in Galileo's *Dialogo sopra i due massimi sistemi del mondo* (1632): 'To me, a great ineptitude exists on the part of those who would have it that God made the universe more in proportion to the small capacity of their reason than to His immense, His infinite, power.'[48] Descartes also emphasized that humans were not the measure of a world they did not create. Humans could not penetrate the reservoir of possible being that was accessible to God's infinite power, nor could they inquire into the reasons for his choices.[49]

In *Syntagma philosophicum* (1658), Gassendi made a similar point. Gassendi wrote that humans tended to glorify themselves as the end of all things. In doing so, however, humans only demeaned the marvellous work of God by adapting it to their own understanding (*ad sui captum*). According to Gassendi, it was a manifestation of self-love when humans thought that the world was created for their sake and their

[43] Descartes, letter to Chanut, 6 June 1647, Descartes 1996, vol. 5, 53–54, cf. Mehl 2013.

[44] For further details see Osler 1994; Beyssade 2000.

[45] Blumenberg 1996b, 194; English translation in Blumenberg, 1985, 173.

[46] A good overview of the long debate on the connection between the emergence of modern science and the irreducibility of divine power can be found in the discussion of Peter Harrison and John Henry, see Harrison 2002 and 2009; Henry 2009, cf. Ashworth 2003.

[47] Augustinus 1998, 34: *Qui ergo dicit: Quare fecit Deus coelum et terram? respondendum est ei, Quia voluit.*

[48] Galilei 1897, 397; English translation in Galilei 2001, 429; cf. Bucciantini 2009.

[49] Descartes 1996, vol. 8, 80–81; English translation in Descartes 1984–1991, vol. 1, 248.

needs (Gassendi 1658, vol. 1, 592a). In fact, God created everything, this whole world, rather for his own sake (*propter semetipsum*) or for his own glory (Gassendi 1658, vol. 1, 151b).

In texts of cited natural philosophers, the emphasis on the divine will invalidated the epistemological openness inherent in the lawful, harmonious, and purposeful cosmos of Renaissance metaphysics and cosmology. In the mechanistic metaphysics, the universe was a decree of divine power and a fact that could be accepted and described, but not rationally grasped or justified.

Natural philosophers further emphasized that humans could not know God's intentions for creation. Descartes put it simply: 'we cannot know God's purposes (*fines*) unless God reveals them'.[50] In his *Dialogue*, Galileo stated that 'it is brash for our feebleness to attempt to judge the reason for God's actions, and to call everything in the universe vain and superfluous which does not serve us'.[51] Humans could never be sure of God's intentions in creation. The universe might contain bodies that humans were unable to see or that served a purpose unimaginable to them. But that was precisely why humans could not let their hubristic ignorance pass judgement on God's works. The same argument was repeated by John Wilkins, Gassendi, or Jean-Baptiste Du Hamel.[52]

The irreducibility of divine power and will meant that human criteria for order, greatness, and beauty were not obligatory for God. For example, human criteria for greatness could not be applied to the universe. Galileo argued that considering the majesty of God, it was entirely adequate to imagine the universe larger than the reach of human cognitive organs. There was an unbridgeable gulf between the extent of human cognition and the extent of creative power.[53] Gassendi and Huygens formulated the same argument later. The greatness of the world was so immense that it exceeded any human measure.[54]

In addition to human notions of greatness, the new natural philosophy questioned other human standards and norms. According to Galileo, even human concepts such as 'large', 'small', 'immense', and 'minute' were not absolute but relative. It was only human conceit to think that they captured the true proportions of things (Galilei 1897, 396). In his *The Usefulness of Experimental Natural Philosophy* (1664) Boyle also wondered how people could think that an omniscient God would create nature to be grasped 'by the dim reason of man'. How could humans think that God 'who may be supposed to operate according to the dictates of his own immense wisdom, should, in his creating of things, have respect to the measure and ease of the human

[50]Descartes, letter to Hyperaspites, August 1641, Descartes 1996, vol. 3, 431; English translation in Descartes 1984–1991, vol. 3, 195.

[51]Galilei 1897, 394, English translation in Galilei 2001, 427.

[52]Wilkins 1802a, 202; Gassendi 1658, vol. 1, 151a, 529a; vol. 4. 57a–b; Du Hamel 1660, 148. A similar argument was already formulated by Galilei, see Galilei 1890–1909, vol. 6, 530.

[53]Galilei 1897, 394; cf. Hamou 2008.

[54]Gassendi 1658, vol. 1, 151, similarly Huygens 1888–1950, vol. 21, 371; cf. Del Prete 2000.

understanding; and not rather, if of any, of angelical intellects?' (Boyle 1775, vol. 2, 46).

Other example of incomprehensibility of God's absolute power concerned the structure of the universe. Gassendi emphasized that humans did not know the nature of heavenly phenomena because the Creator had arranged all this according to his own will (*haec omnia pro arbitrio suo constituerit*). We did not know why the planets moved as we observed them – we only knew that God willed it that way (*sic se habere voluerit*) (Gassendi 1658, vol. 1, 640b). Similar remarks on the unintelligibility of God' decisions can be found in the philosophical reflections of Newton[55] and Huygens.[56]

Kepler knew why the planets were arranged the way we observed them. They were the realization of a mathematical plan hidden in the mind of God, which was accessible to the human mind because in the recognition of mathematical relationships the human mind resembled the divine. Kepler's Platonic belief in the intelligible rationality of the cosmos was rejected by seventeenth-century astronomers and philosophers.

According to Galileo, in creating the world, God disregarded human symmetries and arranged the celestial motions according to ratios that were incomprehensible to human intellect.[57] In *Syntagma philosophicum,* Gassendi stated directly in reference to Kepler that humans were not involved in the divine decisions concerning the arrangement of the stars (Gassendi 1658, vol. 1, 578b; cf. 557a, 567a–b). Humans did not know the true nature of celestial phenomena because the Creator had arranged all this according to his own will. The creating and knowing intellects did not intersect. They were separated by the boundary of transcendence (Gassendi 1658, vol. 1, 640b, 570a–571b).

In his *Astronomia reformata* (1665), the Jesuit astronomer Giovanni Battista Riccioli formulated the same objection against Kepler. Referring to Kepler's hypothesis of Platonic solids, Riccioli wrote that 'God did not take into account proportions which seem to us beautiful' but took into account those which seemed to him suitable for the realization of the ends which he set himself to accomplish in the world (Riccioli 1665, fol. viii–v). Even Riccioli, who otherwise defended Tycho's geoheliocentric cosmology, did not allow for the full accessibility of natural order to human knowledge.

The contrast between the rationalist Platonic conceptions of the cosmos and early modern advocates of its contingency was heightened by new phenomena such as Saturn's ring. In late 1610, Galileo published an anagram in which he cryptically announced observing a strange object around Saturn. Kepler deciphered the anagram incorrectly because he thought that Galileo was announcing the discovery of two

[55]Newton, letter to Bentley, 10 December 1692, in Newton 1961, 234–235; Newton 1726, 528–530; cf. Snobelen 2001; De Smet and Verelst 2001; Henry and McGuire 2018.

[56]Huygens 1888–1950, vol. 21, 514–516, 556; see also Mormino 2000.

[57]Galilei, letter to Gallanzone Gallanzoni, 16 July 1611; in Galilei 1890–1909, vol. 11: 149; cited by Palmerino 2006, 37.

moons of Mars. Kepler's error came from his belief that the placement of bodies in the cosmos could not be contingent. His speculative mathematical vision of the cosmos led him to the idea that there must be two moons for Mars and four for Jupiter, because the ratio of the moons of Earth, Mars, Jupiter, and Saturn should be 1, 2, 4, and 8.[58] Kepler's mistake nicely illustrated that the Platonic belief in affinity between human and divine minds excluded contingency and preserved the notion of the world as a rational order, thus ensuring the possibility of its *a priori* rational comprehension.

Huygens seemed to share this opinion in his early research. When he discovered Saturn's moon Titan in 1655, he ventured to assert (*confirmare audeam*) that no more moons would be discovered in the future because the number of bodies in the system was thus fixed at twelve, there being six planets and six moons in the system (Huygens 1888–1950, vol. 15, 213–214). As Bernard I. Cohen demonstrated, this number was considered perfect by the young Huygens, following the numerological tradition (Cohen 1978). In his later age, Huygens changed his mind. In his last treatise, *Cosmotheoros* (1698), Huygens argued that telescopes revealed an immeasurable number of stars, and more would probably show themselves to human eyes. Nothing prevented the possibility that, beyond the last visible stars, God created an infinite number of other things that completely escaped our understanding (*res alias innumeras a cogitationibus nostris remotas*) (Huygens 1888–1950, vol. 21, 815–817, cf. 371).

I mentioned that Renaissance philosophers sometimes justified the nobility of humans by saying that they were created as an epilogue of nature. Galileo reversed this story. In his view, God did not intend an inherent affinity between nature and man, and therefore separated their creation. Galileo criticized Aristotelians who resorted to the anthropocentric claim that our comprehension was adapted to understand the world 'as if nature had first made the brain of man, and then arranged everything to conform to the capacity of his intellect. But I should think rather that nature first made things in her own way, and then made human reason skilful enough to be able to understand, but only by hard work, some part of her secrets.'[59] Galileo thus suggested, as Philippe Hamou aptly noted, that human science was characterized by a certain epiphenomenalism: it grafted on a reality that ignored, preceded, and transcended it (Hamou 2008, 165).

Later in the century, Boyle also argued against the same idea in a way similar to Galileo: 'the world itself was first made before the contemplator of it, man: whence we may learn, that the author of nature consulted not, in the production of things, with human capacities, but first mad things in such manner as he was pleased to think fit, and afterwards left human understanding to speculate…' (Boyle 1775, vol. 5, 694; cf. vol. 6, 794). The Renaissance cosmos had turned into a mechanistic universe whose arrangement appeared entirely contingent and incomprehensible to humans.

---

[58] Kepler 1938–2009, vol. 4, 309; on Saturn see Kepler 1938–2009, vol. 4, 291; cf. Hallyn 1992, 94.

[59] Galilei 1897, 289; English translation in Galilei 2001, 307.

Renaissance natural philosophy inspired by Platonism assumed that there was a symmetry between human curiosity and nature that provided answers to an unchanging set of human questions. Representatives of early modern mechanistic natural philosophy believed that God did not give man insight into the purpose and meaning of various parts of the world. Humans lacked insight into the divine mind and the reasons that guided the will of God in creating the natural world. The Renaissance symmetry between human curiosity and nature that could reflect itself in human knowledge was abolished and abandoned. Things did not bear signatures, signs of their meaning, and significance for humans. Therefore, human knowledge must have been temporal, slow, gradual, limited, unfinished, and facing unexpected and unpredictable natural phenomena. Mechanism brought about a substantial reconfiguration of the ontological and epistemological tenets of natural philosophy that opened the possibility of the idea of scientific progress.

2.4 Enlargement of the Universe: Cosmological Contingency

Another factor that significantly challenged the notion of the anthropocentric nature of the cosmos was further advances in astronomy. In this case, the rejection of anthropocentrism did not come from metaphysical, ontological, and epistemological reasoning, but from empirical findings that were further speculated upon. The result was a picture of the universe conveying a vastness beyond all human expectations. In recent research, much attention has been paid to the idea of an infinite universe (Seidengart 2006; Del Prete 1999, 2000), but only a few authors have addressed the unsettling numerical parameters of the universe introduced by the new astronomy (Van Helden 1985, 118–159).

Ptolemy assumed that all the stars were the same distance away because they were all embedded in the same spherical shell surrounding the cosmos. He guessed the distance from Earth to stars at approximately 20,000[60] earth radii (ER).[61] Most astronomers from late antiquity to the early seventeenth century, with various variations, adhered to Ptolemy's values for the distances of stars. Tycho reduced this distance to 14,000 ER.[62] During the seventeenth century, estimates of the distances of celestial bodies were refined, and the actual values turned out to be much larger than Ptolemaic ones. Astronomers attempted to calculate the distance of

[60] See Ptolemy 1967, 7; cf. Evans 1998, 68–74, 387–389.

[61] Earth radius – a unit commonly used in the history of astronomy to express distances in space; Riccioli presented a comprehensive table of estimates of the size of Earth by various astronomers, himself giving the overly large value of ER 5174 Roman miles, i.e. 7674 km; for the German astronomers Riccioli gave a value of 860 German miles, i.e. about 6450 km (Riccioli 1651, vol. 1, 63).

[62] Brahe 1913–1929, vol. 2, 430; cf. Lerner 2008, vol. 2, 109–110.

celestial bodies by estimating the value of their parallax. However, reliable measurements of parallax were beyond the technical capabilities of seventeenth-century astronomers (the first measurement of stellar parallax was not made until the late 1830s), so they either gave made-up values or explicitly pointed out that these were very rough estimates.[63]

Tycho estimated that if heliocentric cosmology were correct, third-magnitude stars would be 7,850,000 ER away and first-magnitude stars would have to be even more distant (Brahe 1913–1929, vol. 6: 197). Tycho's fears soon proved true, for Kepler presented values that must have shocked his contemporaries. In *De stella nova* (1606) he stated that the distance to the sphere of the fixed stars was 34,077,066 ER (Kepler 1938–2009, vol. 1, 235). In his later work, *Epitome astronomiae copernicanae,* Kepler estimated the distance to the stars at 60,000,000 ER (see Lerner 2008, vol. 2, 128–129).

A good source of information on estimating the distances of stars in the mid-seventeenth century was the extensive work of Riccioli, *Almagestum novum* (1651). The sixth book of this work contained tables in which Riccioli gave values of the minimum distances of stars according to various astronomers. For non-Copernican authors, Riccioli usually gave the Ptolemaic number of 20,000 ER (Riccioli 1651, vol. 1, fol 419a–b, 680–683). Riccioli himself gave 210,000 ER as his estimate of the distances of the stars. The values for the size of the heliocentric (closed) universe in his table varied between 13 million ER (Galileo) and 60 million ER (Kepler). From Riccioli's tables alone, it was clear that the Copernican universe was at least a thousand times larger than the geocentric universe. Yet all of these values were still too low. We know that the nearest star, Proxima Centauri, is located at a distance of 6.3 billion ER.

In the second half of the seventeenth century, astronomers could rely on better telescopes and new observational tools such as the micrometre. They were thus able to determine the distance to the Sun with relative accuracy, from which the distances to the planets in the solar system were calculated. In his *Systema Saturnium* (1659), Huygens presented new estimates for the distances and sizes of the planets. Huygens's values were already close to those we know. Huygens estimated the Sun-Earth distance fairly accurately at 25,086 ER. He also estimated the minimum distance of Saturn at 200,000 ER (Huygens 1888–1950, vol. 15, 347–349). After constructing and using his tubeless telescope in the 1680s, Huygens made some measurements, and in his late work *Cosmotheoros* he further speculated on the distances of stars. He located Sirius at a distance of about 660,000,000 ER (Huygens 1888–1950, vol. 21, 815–817).

Newton supposed the universe to be even bigger. In his posthumously published *A Treatise of the System of the World,* he wrote that the nearest star must be a hundred thousand times farther away than the distance of Saturn from the Sun, for which he gave Huygens' value of 200,000 ER (Newton 1731, 91). In sum, at the

[63] Riccioli 1665, fol. viii–a; Gassendi 1658, vol. 1, 150b, 580a; Guericke 1672, 224. See excellent overview in Siebert 2005 and more detailed discussion in Siebert 2006, 155–293.

beginning of the seventeenth century, Ptolemy's guess of 20,000 ER was still commonly assumed. By the end of the century, distances of 20 billion ER were already being considered.

It was clear to astronomers and philosophers that their readers could not easily imagine such huge numbers. Therefore they converted the distances into temporal values for the fall of heavy bodies. Athanasius Kircher claimed that a millstone would fall from the stars to Earth in six hours (Kircher 1660, 357). Riccioli again prepared a table in which he calculated how much time, according to the values of individual astronomers, an iron ball weighing one pound would take to fall from the Sun to Earth. Riccioli's table gave mostly values of one or two days. Another table indicated that it would take about 100 days for the ball to fall from the stars (Riccioli 1651, vol. 1, 697).

These comparisons made their way into contemporary literature. In his *Les Caractères* (1688), Jean da la Bruyère stated that if a millstone were to fall from the Sun towards Earth, it would take 114 years for it to fall. Saturn, however, was so distant that the stone would take 1140 years to fall. Saturn's orbit around the Sun was so long that an English horse would not have traversed it before 20,548 years (Bruyère 1985, 449–450). Huygens abandoned the idea of falling bodies and introduced another analogy: A ball fired from a cannon would take 25 years to voyage from Earth to the Sun. From Saturn, it would take 250 years to travel to the Sun (Huygens 1888–1950, vol. 21, 807). And a cannonball would fly 700,000 years to the nearest stars (Huygens 1888–1950, vol. 21, 815–817).

Parallel to the empirical enlargement of the universe, the conceptual minimization of Earth was taking place. In the vast universe, Earth was turning into a negligible and ephemeral entity. Astronomers and natural philosophers have often referred to the fact that, because of the immeasurable parallax, Earth was like a point in relation to the planets and stars (Huygens 1888–1950, vol. 21, 747, 807).

The analogy of the imperceptible point was then applied to the scope of human empirical knowledge. Some natural philosophers asserted human knowledge to be a mere point in comparison to the extent of the universe. For example, Boyle wrote (Boyle 1775, vol. 4, 52):

> the earth is but a physical point in comparison of the starry heaven: of how little extent must our knowledge be, which leaves us ignorant of so many things touching the vast bodies, that are above us, and penetrate so little a way even into the earth, that is beneath us, that it seems confined to but a small share of the superficial part of a physical point.

This topos of the pointlike nature of human knowledge appeared in early modern natural philosophical literature in numerous variations[64] always with the same message: human empirical knowledge covered only a tiny part of the universe. At the end of his life, Newton expressed this feeling in a famous metaphor: 'I seem to have been only like a boy playing on the sea shore, and diverting myself in now and

[64] Wilkins 1802b, 101; Borel 1657, 3; Huygens 1888–1950, vol. 21, 368; Leeuwenhoek, letter to Huygens, 20 May 1679, in Leeuwenhoek 1939–1999, vol. 3, 66; Locke 1975, 545–547; Fontenelle 1990–2001, vol. 6, 47; Pascal 1998, vol. 2, 608–610; Régis 1690, vol. 2, 108.

then finding a smoother pebble or a prettier shell than ordinary, whilst the great ocean of truth lay all undiscovered before me' (Westfall 1981, 863). It is significant that Newton referred to his research as a child's game, as if true science were just beginning and yielding its first immature, limited, and fragmentary results. The metaphor of the immense ocean indicated the vastness of what remained to be discovered. The vast, unknown and contingent universe was another condition of possibility for the birth of the idea of scientific progress because the perfectioning of knowledge was thinkable only in the world where new discoveries could be expected.

2.5 The Plurality of Worlds: Anthropological Contingency

Mechanistic metaphysics introduced the belief that humans could not understand the architectural decisions of the Creator but this did not mean that God had become irrelevant to natural philosophy and sciences. The emphasis on the causal impotence of nature and its inability to self-organize turned into an argument from design: the order and variety observed in the world could only be the creation of an intelligent creator. Many natural philosophers shared Boyle's belief that the universe and its inhabitants were not the result of 'blind chance' or 'tumultuous justlings of senseless matter' (Boyle 1775, vol. 5, 514). This view was repeated by Cartesians[65] and British natural philosophers, such as Roger Cotes in his introduction to the second edition of Newton's *Principia* (Cotes 2004, 58, cf. Newton 1726, xxxii). Mechanism thus contributed to the flowering of physicotheology at the end of the seventeenth century. Some historians even speak about 'the physicotheological functionalization of the mechanical world-picture as a proof of God.'[66]

The contingency of nature, therefore, did not completely exclude teleology in the mechanistic paradigm.[67] But the function of teleology has changed compared to Renaissance natural philosophy. Teleology was no longer to be an explanatory and intelligible principle of natural phenomena; its function was to foster piety. The other difference to the Renaissance concept of purposefullness concerned the focus of teleology. In the mechanistic paradigm, nature could retain some traces of teleology, but its teleology was not exclusively human-centred.

The idea of plurality of inhabited worlds is a good exemplification of this change. It denied cosmological anthropocentrism: human were not at the centre of the world. The idea also excluded anthropofinalist assumption that the stars shone only because of man. Gassendi assumed that stars were not created for the sake of humans. God placed the stars in the far reaches of the universe to illuminate, warm, and enliven places where the light of our sun could not reach (Gassendi 1658, vol. 1, 667b). In

---

[65] Rohault 1671, fol. a3v–a4v; Régis 1690, vol. 1, 271–273; vol. 2, 111; cf. Clarke 1989, 107–154.

[66] Sheehan 2003, 50; cf. Shanahan 1994, 191; Groh and Groh 1991, 50.

[67] See more closely Des Chene 2000a; Schmid 2010.

Guericke's view, the stars shone to illuminate their planets so that their light was not useless (*ne lux earum sit frustranea*) (Guericke 1672, 231a). In his physico-theological treatise, John Ray succinctly wrote: 'For it is highly absurd and unreasonable, to think that Bodies of such vast Magnitude as the fix'd Stars were only made to twinkle to us' (Ray 1735, 178–179). A similar view was shared by other representatives of British physico-theology.[68]

Thus, at the end of the century, the belief that the universe and nature served beings other than human became widespread. This idea was justified in two ways. The first one was belief in the economy and expediency of God's work. There must be other inhabitants of the cosmos. Otherwise, much of God's creation would exist in vain. Stars shone to provide the conditions for life for extraterrestrial beings who existed to bear witness to the wisdom of the Creator.[69] In his *Discours nouveau prouvant la pluralité des Mondes* (1657), Pierre Borel wrote that it was hard to believe that the stars would move only for the benefit of one planet, which was the most sordid (*le plus vil*) and almost the smallest of all. God, tired of loneliness, created the universe as an empire whose inhabitants admired and worshipped him.[70]

Apart from the physico-theological justification, the idea of a plurality of inhabited worlds was introduced by other reflection: the impossibility of privileging Earth. In the pluralistic universe of the second half of the seventeenth century, the principle of the terrestrial analogy prevailed (Siebert 2006, 54–55). According to the principle, it was reasonable to assume that what existed on Earth was also found on other planets including life and the presence of rational beings. This principle and the related themes of the plurality of inhabited worlds and extraterrestrial life were most fully developed in two works that appeared in the late seventeenth century.

The first was published in 1686 by Fontenelle under the title *Entretiens sur la pluralité des mondes*. The work took the form of a conversation between a philosopher and an inquisitive marquise. In a series of six dialogues, the philosopher revealed to the marquise the secrets of the universe by defending heliocentrism and explaining Cartesian vortex cosmology.

The other treatise was different. While Fontenelle was only at the beginning of his philosophical career in 1686, *Cosmotheoros* (published in 1698) represented the last work of the famous scientist Huygens. The Dutch scientist had worked on the book since the 1680s and he himself prepared the work for publication, but the volume was not published until after his death (Mormino 2000, 502; Schoot 2014).

Both works were exceptionally popular at the turn of the seventeenth and eighteenth centuries. They were published in numerous editions and translated into other European languages.[71] Their cosmological content, astronomical argumentation, and literary strategy have already been presented several times in recent

[68] Bentley 1838, 175; Derham 1738, xlvii, 40–41; Derham 1724, 111, 165, 282–288; cf. Matytsin 2013; Fara 2004; Mandelbrote 2007; Armogathe 2007, 251–263.

[69] Guthke 1983, 91–105; Dick 1982, 69–101; Darmon 2002.

[70] Borel 1657, 14; cf. Rheita 1645, 179b; Guericke 1672, 216a.

[71] The history of the text of *Entretiens* was described in detail by Poulouin 2013, 98–118.

research.[72] From my point of view both writings are important because they aptly illustrate the reconfiguration of the human position in the universe which resulted in different evaluation of conditions of knowledge.

Huygens and Fontenelle derived the existence of life on other planets from the denial of all three variants of anthropocentrism. In their view, humans vainly and conceitedly thought that their planet was the only inhabited and animated place in the cosmos. Earth could not be a cosmic exception. If elements, plants, and animals occurred on other planets, they could not be denied that which constituted the 'ornament' of every planet – rational beings, *planeticoli.* [73]

Instead of a simple universal teleology, according to which everything served the needs of man, both authors advocated a teleologically multifocal universe with countless worlds populated by rational beings for whom the universe was created. The human race represented only one kind of these beings (Matytsin 2013).

Jean-François Stoffel demonstrated that the advocates of heliocentrism in the seventeenth century created a compelling and persuasive rhetorical figure that identified the origins of geocentrism in anthropocentrism without taking into account the astronomical and physical reasons of the advocates of heliocentric astronomy (Stoffel 1998, 2001, 2002). In this way, geocentrism was to be disqualified as an irrational manifestation of human pride, pampering, and self-love, as found for instance in Cyrano de Bergerac and Huygens.[74] In their view, childish cosmological anthropocentrism resulted in geocentric cosmology.

In his *Entretiens*, Fontenelle further worked out this strategy. At the beginning of the dialogue, Fontenelle connected the anthropocentric concept of the universe with the historical development of astronomy. He thus spoke of anthropocentrism in connection with the origins of astronomy among the ancient Chaldeans, who brought the first concept of the universe. According to Fontenelle, however, human nature was immutable, and the same psychological mechanism that brought the Chaldeans to adopt anthropocentrism and geocentrism also appeared in contemporary humans, unless they were enlightened by modern natural philosophy:[75]

> Our folly is to believe that all of nature, without exception, is destined for our use, and when one inquires of the philosophers what is the use of the prodigious number of fixed stars, when a fraction would accomplish the same thing, they answer coldly that they serve to please our sight.

In Fontenelle's view, geocentrism was thus a misguided notion based on universal human prejudices of self-importance and self-love. Both Fontenelle and Huygens understood such a conception of the universe as the manifestation of the rational maturity and emancipation that earlier humans had been incapable of because they

[72] Aït-Touati 2011, 91–125; Seidengart 2006, 537–560; Lambert 2002, 125–139; Dick 1982, 124–136; Guthke 1983, 202–216; Knobloch 2002; Radelet de Grave 2003.

[73] Fontenelle 1991, 69–71; Huygens 1888–1950, vol. 21, 709–715; cf. Rossi 1972.

[74] Cyrano de Bergerac 1921, *Les estats de la Lune*, 13–14; cf. similarly Huygens 1888–1950, vol. 21, 795.

[75] Fontenelle 1991, 24; English translation in Fontenelle 1990, 14.

were trapped in their childish self-centered prejudices. Fontenelle further developed his history of astronomy to Copernicus whom he saw as a philosopher who had revealed the bitter truth to mankind: 'He did well [. . .] to have put down the vanity of men, who had given themselves the greatest place in the universe.'[76] The vanity and stupidity of humans were universal. In Fontenelle's view, even the ancient sage put himself at the centre of the world and thought that everything was made for him and 'unconsciously accepts that principle which flatters him'.[77]

Isabelle Mullet pointed out that the combination of geocentrism and self-love took on a psychological depth in *Entretiens* when Fontenelle connected the revolutions of the heavens with the movements of the human mind (Mullet 2011, 71). The heliocentric order of the heavenly bodies was not directly perceptible. Similarly, the most natural movements of the mind were also difficult to comprehend: 'The working of self-love is so natural in us that usually we don't even sense it, and believe we're acting on other principles.'[78] An important conclusion followed this short reflection. If an untruth was a manifestation of human self-love, then truth was painful and went against the grain of humans. The more a given theory discredited the importance of humans in the order of nature, the closer it would be to the truth, because such a theory was presumably untainted by the egocentricity of the human race.

We can see this clearly, for example, in the way Huygens described both terrestrial and extraterrestrial science. Development of natural knowledge, either on Earth or on the planets, brought natural philosophers to the acknowledgement of their own cosmological insignificance (e. g. Huygens 1888–1950, vol. 21, 823). The anthropological proof of the truth of heliocentrism, then, was ultimately to be precisely its indifference to humans. Fontenelle and his contemporaries assumed that a theory unflattering to man was closer to the truth because it required an objective self-denial on the side of humans, a rationally enforced rejection of intuitive ideas imposed on human consciousness by uncritical self-love. The natural world no longer seemed to be a source of stimuli awakening the seeds of God's wisdom in human souls.

The anthropocentric view was thus identified with intellectual dependency. The ability to admit the bitter truth was interpreted as a sign of intellectual maturity and development. According to Blumenberg, the self-confidence of early modern philosophy derived from transforming the cosmic eccentricity of man into a positive vision of a humanity that could take care of itself: 'the self-consciousness of modern man was to be based on the negation of the possibility that he could read off from nature, as the Creation, whether any role at all had been "intended" for him and what

[76]Fontenelle 1991, 28; English translation in Fontenelle 1990, 16.

[77]Fontenelle 1991, 28; English translation in Fontenelle 1990, 17.

[78]Fontenelle 1991, 36; English translation in Fontenelle 1990, 22.

role it was.'[79] Humans started to be considered beings who themselves decided on their position in the universe on the basis of accumulated knowledge.

The unmasking of anthropocentric teleology as a prejudice originating at the beginning of history was another condition for the emergence of the idea of scientific progress. Natural philosophy came to understand itself as an emancipatory force that liberated humans from infantile delusion. This new self-understanding of natural philosophy naturally paved the way for the Enlightenment identification of scientific progress as the main force in the history of humankind.

Mechanistic ontology, voluntarist theology, and astronomical measurements accompanied by cosmological speculation brought natural philosophers to believe that humans could know neither the natural world, nor its essence, nor its fundamental principles, nor its full extent. Human criteria were not binding on the Creator, human explanations of natural processes were mere artificial constructions, and human experience was limited to only a tiny part of the immense universe. This unpleasant epistemological conclusion was mitigated by the introduction of time into knowledge. The idea of scientific progress compensated for new epistemological findings by insisting on the temporariness of human ignorance and promising further discoveries in the boundless ocean of truth.

Bibliography

Agrippa of Nettesheim, Cornelius. 1533. *De occulta philosophia libri tres*. Cologniae: Publisher unknown.

Aït-Touati, Fréderique. 2011. *Fictions of the cosmos. Science and literature in the seventeenth century*. Chicago/London: The University of Chicago Press.

Armogathe, Jean-Robert. 2007. *La nature du monde. Science nouvelle et exégese au XVIIe siécle*. Paris: Presses Universitaires de France.

Ashworth, William B. 2003. Christianity and the mechanistic universe. In *When science & Christianity meet*, ed. David Charles Lindberg – Ronald Leslie Numbers, 61-84. Chicago/London University of Chicago Press.

Augustine of Hippo. 1998. *De Genesi contra Manichaeos*. In *Corpus Scriptorum Ecclesiasticorum Latinorum*, ed. Dorothea Weber, vol. 91. Wien: Österreichische Akademie der Wissenschaften.

Barker, Peter. 1997. Kepler's epistemology. In *Method and order in Renaissance philosophy*, ed. Daniel Antonio Di Liscia, Eckhard Kessler, and Charlotte Methuen, 355–368. Aldershot: Ashgate.

Barker, Peter, and Bernard R. Goldstein. 2001. Theological foundations of Kepler's astronomy. *Osiris* 16: 88–113.

Bentley, Richard. 1838. *Eight Boyle lectures: A confutation of atheism from the origin and frame of the world*. In *The works of Richard Bentley*, ed. Alexander Ryce, vol. 3, 119–200. London: Robson.

Beyssade, Jean-Marie. 2000. Toute-pouissance de Dieu et nécessité des principes physiques chez Descartes. In *Potentia Dei. L'omnipotenza nel pensiero dei secoli XVI e XVII*, ed. Guido Canziani, Miguel Ángel Granada, and Yves Charles Zarka, 351–368. Milan: Franco Angeli.

[79] Blumenberg 1996a, 420; English translation in Blumenberg 1987, 356.

Bianchi, Massimo L. 1994. The visible and invisible from alchemy to Paracelsus. In *Alchemy and chemistry in the 16th and 17th centuries*, ed. Pioy Rattansi and Antonio Clericuzio, 17–50. Dordrecht: Kluwer.

———. 2004. Signs, signature, and natursprache in Paracelsus and Böhme. In *Res et Verba in der Renaissance*, ed. Eberhard Kessler and Ian McLean, 197–215. Wiesbaden: Harrassowitz.

Blair, Ann. 1997. *The theatre of nature. Jean Bodin and Renaissance science*. Princeton: Princeton University Press.

Blumenberg, Hans. 1970. *Selbsterhaltung und Beharrung: Zur Konstitution der neuzeitlichen Rationalität*. Wiesbaden: Harrassowitz.

———. 1985. *The legitimacy of the modern age*, translated by Robert M. Wallace. Cambridge: MIT Press.

———. 1987. *The genesis of the Copernican world*, translated by Robert M. Wallace. Cambridge: MIT Press.

———. 1996a. *Die Genesis der kopernikanischen Welt*. Frankfurt am Main: Suhrkamp.

———. 1996b. *Die Legitimität der Neuzeit. Erneuerte Ausgabe*. Frankfurt am Main: Suhrkamp.

Bodin, Jean. 1597. *Universae naturae theatrum*. Francofurti: Wechelius.

Borel, Pierre. 1657. *Discours nouveau prouvant la pluralité des Mondes*. Genève: Publisher anonymous.

Boyle, Robert. 1775. The works, ed. Thomas Birch. 6 vols. London: Rivington.

Brahe, Tycho. 1913–1929. *Tychonis Brahe Dani opera omnia*, ed. John L. Dreyer, 10 vols. Copenhagen: Gyldendal.

Bruyère, Jean de la. 1985. *Caractères*, ed. Pierre Ronzeaud. Paris: Livre de Poche.

Bucciantini, Massimo. 2008. *Galilée et Kepler: Philosophie, cosmologie et théologie à l'époque de la Contre-Réforme*. Paris: Les Belles Lettres.

———. 2009. Descartes, Mersenne et la Philosophie Invisible de Galilée. *Dix-septième siècle* 61: 19–30.

Cardano, Girolamo. 1663. *Opera omnia*. 10 vols. Lugduni: Huguetan.

Cassirer, Ernst. 1922. *Das Erkenntnisproblem in der Philosophie und Wissenschaft der neueren Zeit*. 3 vols. Berlin: Olms.

Clarke, Desmond. 1989. *Occult powers and hypotheses. Cartesian natural philosophy under Louis XIV*. Oxford: Oxford University Press.

———. 2018. *French philosophy, 1572–1675*. Oxford: Oxford University Press.

Cohen, Bernard I. 1978. Perfect numbers in the Copernican system: Rheticus and Huygens. *Studia Copernicana* 16: 419–425.

Comenius, Johannes A. 1964. *De rerum humanarum emendatione consultatio catholica*. 2 vols. Academia: Pragae.

Copernicus, Nicolaus. 1975. De revolutionibus. In Nicolaus Copernicus, *Opera omnia*, ed. Ryszard Gansiniec, vol. 2. Warsaw/Cracow: Officina publica libris scientificis edendis.

———. 1978. On the revolutions. In Nicolaus Copernicus, *Complete works*, ed. Jerzy Dobrzycki, translated by Edward Rosen. Warsaw/Cracow: Polish Scientific Publishers.

Cotes, Roger. 2004. Editor's preface to the 2nd edition of *Philosophia naturalis principia mathematica* (1713). In Isaac Newton, *Philosophical writings*, ed. Andrew Janiak, 42–59. Cambridge: Cambridge University Press.

Croll, Oswald. 1620a. *Basilica Chymica*. Francofurti: Tampachius.

———. 1620b. De signaturis rerum. In Oswald Croll, *Basilica Chymica*. Francofurti: Tampachius (separated pagination).

———. 1670. *Royal and practical chymistry in three treatises*. London: Starkey.

Cyrano de Bergerac. 1921. *Les estats et empires de la lune*. In Cyrano de Bergerac, *Les œuvres libertines*, ed. Frédéric Lachèvres. Paris: Champion.

Darmon, Jean-Charles. 2002. L'imagination de l'espace entre argumentation philosophique et fictions. *Études Littéraires* 34: 217–239.

Daston, Lorraine. 1995. How nature became the other: Anthropomorphism and anthropocentrism in early modern natural philosophy. In *Biology as society, society as biology*, ed. Sabine Maasen and Everett Mendelsohn, 37–56. Dordrecht: Springer.
———. 2000. Historische Überlegungen zum Antropomorphismus und zur Objektivität in den Wissenschaften. In *Die Einheit der Wirklichkeit: zum Wissenschaftsverständnis der Gegenwart*, ed. Bernd-Olaf Küppers, 27–44. München: Fink.
De Smet, Rudolf, and Karin Verelst. 2001. Newton's scholium generale: The Platonic and Stoic legacy – Philo, Justus Lipsius and the Cambridge Platonists. *History of Science* 39: 1–30.
Del Prete, Antonella. 1999. *Bruno, l'Infini et les mondes*. Paris: Presses Universitaires de France.
———. 2000. Pierre Gassendi et l'univers infini. In *Libertinage et philosophie au XVIIe siècle*. Tome 4: *Gassendi et les gassendistes*, ed. Antony McKenna – Pierre-François Moreau, 57–69. Saint-Étienne: Publications de l'Université de Saint-Étienne.
Derham, William. 1724. *Physicotheology. Demonstration of the being and attributes of God from his works of creation*. London: Innys – Richardson.
———. 1738. *Astro-theology: Or, a demonstration of the being and attributes of God, from a survey of the heavens*. London: Innys/Manby.
Des Chene, Dennis. 2000a. *Physiologia: Natural philosophy in late Aristotelianism*. Ithaca: Cornell University Press.
———. 2000b. *Spirits and clocks. Machine and organism in Descartes*. Ithaca: Cornell University Press.
———. 2005. Mechanisms of life in the seventeenth century: Borreli, Perrault, Régis. *Studies in History and Philosophy of Biological and Biomedical Sciences* 36: 245–260.
Descartes, René. 1984–1991. *The philosophical writings*. 3 vols, translated by John Cottingham, Robert Stoothoff, Dugald Murdoch, and Athony Kenny. Cambridge: Cambridge University Press.
———. 1996. *Œuvres de Descartes*, ed. Charles Adam and Paul Tannery, 11 vols. Paris: Vrin.
Dick, Steven J. 1982. *Plurality of worlds. Origins of the extraterrestrial life debate from Democritus to Kant*. Cambridge: Cambridge University Press.
Du Hamel, Jean-Baptiste. 1660. *Astronomia Physica, sedu de luce, natura, et motibus corporum coelestium*. Paris: Lamy.
Evans, James. 1998. *History and practice of ancient astronomy*. Oxford: Oxford University Press.
Fara, Patricia. 2004. Heavenly bodies: Newtonianism, natural theology and the plurality of worlds debate in the eighteenth century. *Journal for the History of Astronomy* 35: 143–160.
Ficino, Marsilio. 1576. *Opera omnia*. Basileae: Officina Henricpetrina.
———. 2001–2006. *Platonic theology*, ed. James Haskins, 6 vols. Cambridge, MA/London: Harvard University Press.
———. 2002. *Three books on life*, ed. V. Carol, John Kaske, and R. Clark. Tempe, Arizona: Arizona Center for Medieval and Renaissance Studies.
Fontenelle, Bernard de. 1990. *Conversations on the plurality of worlds*, translated by H. A. Hargreaves. Berkeley: University of California Press.
———. 1990–2001. *Œuvres complètes*, ed. Alain Niderst. Paris: Fayard.
———. 1991. *Entretiens sur la pluralité des mondes*. In Bernard de Fontenelle. *Œuvres complètes*, ed. Alain Niderst, vol. 2, 10–130. Paris: Fayard.
Galilei, Galileo. 1890–1909. *Edizione nazionale delle opere di Galileo Galilei*, ed. Antonio Favaro, 20 vols. Firenze: Barbèra Editore.
———. 1897. *Dialogo sopra i due massimi sistemi del mondo*. In *Edizione nazionale delle opere di Galileo Galilei*, ed. Antonio Favaro, 1–489. Firenze: Barbèra Editore.
———. 2001. *Dialogue concerning the two chief world systems: Ptolemaic and Copernican*, translated by Stillman Drake. New York: The Modern Library.
Garber, Daniel. 1992. *Descartes' metaphysical physics*. Chicago/London: University of Chicago Press.

———. 2001. How God causes motion: Descartes, divine sustenance, and occasionalism. In Daniel Garber, *Descartes embodied: Reading Cartesian philosophy through Cartesian science*, 189–202. Cambridge: Cambridge University Press.
Gassendi, Pierre. 1658. *Opera omnia*. 6 vols. Lugduni: Anisson.
Gaukroger, Stephen. 2006. *The emergence of a scientific culture: Science and the shaping of modernity, 1210–1685*. Oxford: Oxford University Press.
———. 2010. *The collapse of mechanism and the rise of sensibility. Science and shaping of modernity, 1680–1760*. Oxford: Oxford University Press.
———. 2016. *The natural and the human. Science and the shaping of modernity, 1739–1841*. Oxford: Oxford University Press.
Glanvill, Joseph. 1885. *Scepsis scientifica, or vanity of dogmatizing*, ed. James Owen, London: Kegan Paul, Trench & Company.
Granada, Miguel Ángel. 2013. Tycho Brahe's anti-Copernican campaign. In *Celestial novelties on the eve of the scientific revolution*, ed. Dario Tessini and Patrick J. Boner, 185–207. Florence: Olschki.
Groh, Dieter. 2003. *Schöpfung im Widerspruch: Deutungen von der Natur des Menschen von der Genesis bis zur Reformation*. Frankfurt am Main: Suhrkamp.
———. 2010. *Göttliche Weltökonomie: Perspektiven der wissenschaftlichen Revolution von 15. bis zum 17. Jahrhundert*. Frankfurt am Main: Suhrkamp.
Groh, Ruth, and Dieter Groh. 1991. *Weltbild und Naturaneignung. Zur Kulturgeschichte der Natur*. Frankfurt am Main: Suhrkamp.
Guericke, Otto von. 1672. *Experimenta Nova Magdeburgica de Vacuo Spatio*. Amstelodami: Waesberg.
Guthke, Karl S. 1983. *Der Mythos der Neuzeit. Das Thema der Mehrheit der Welten in der Literatur – und Geistesgeschichte von der kopernikanischen Wende bis zur Science Fiction*. Bern/München: Francke.
Hadot, Pierre. 1995. *Philosophy as a way of life. Spiritual exercises from Socrates to Foucault*. Oxford: Oxford University Press.
Hallyn, Fernand. 1992. Introduction. In Galileo Galilei, *Le Messager des étoiles*, 11–130. Paris: Seuil.
Hamou, Philippe. 1999. *La mutation du visible. Essai sur la portée épistemologique des instruments d'optique au XVIIe siècle*. 2 vols. Paris: Septentrion.
———. 2008. La nature est inexorable. Pour une reconsidération de la contribution de Galilée au problème de la connaissance. *Galilaeana* 5: 149–177.
Harrison, Peter. 2002. Voluntarism in early modern science. *History of Science* 40: 63–89.
———. 2009. Voluntarism and the origins of modern science: A reply to John Henry. *History of Science* 47: 223–231.
Hattab, Helen. 2019. Descartes's mechanical but not mechanistic physics. In *The Oxford handbook of Descartes and Cartesianism*, ed. Steven Nadler, Tad M. Schmaltz, and Delphine Antoine-Mahut, 124–137. Oxford: Oxford University Press.
Henry, John. 2009. Voluntarist theology at the origins of modern science: A response to Peter Harrison. *History of Science* 47: 79–113.
Henry, John, and J.E. McGuire. 2018. Voluntarism and panentheism: The sensorium of God and Isaac Newton's theology. *The Seventeenth Century* 33: 587–612.
Hirai, Hiro. 2016. The word of God and the universal medicine in the chemical philosophy of Oswald Croll. In *Alchemy and Rudolf II: Exploring the secrets of nature in central Europe in the 16th and 17th centuries*, ed. Ivo Purš, 381–385. Prague: Artefactum.
———. 2019. Into the Forger's library: The genesis of 'De natura rerum' in publication history. *Early Science and Medicine* 24: 485–503.
Hirai, Hiro, and Yohei Kikuchihara. 2015. Signatura rerum theory. In *Encyclopedia of Renaissance philosophy*. Dordrecht: Springer.
Huygens, Christiaan. 1888–1950. *Œuvres complètes*. 22 vols. La Haye: Nijhoff.

Kepler, Johannes. 1938–2009. *Gesammelte Werke*, ed. M. Caspar, V. Bialas et al., 25 vols. München: C. H. Beck.

———. 1995. *Epitome of Copernican Astronomy. Harmonies of the world*, translated by Charles Glenn Wallis, Prometheus Book.

Kessler, Eberhard. 1992. Selbstorganisation in der Naturphilosohie der Renaissance. *Selbstorganisation. Jahrbuch für Komplexität in der Natur-, Sozial- und Geisteswissenschaften* 3: 15–29.

———. 1994. Naturverständnisse im 15. und 16. Jahrhundert. In *Naturauffassungen in Philosophie, Vol. 2: Renaissance und frühe Neuzeit*, ed. Lother Schäfer and Elisabeth Ströker, 13–57. Freiburg/München: Alber.

Khunrath, Heinrich. 1609. *Amphitheatrum sapientiae aeternae*. Francofurti: Antonius.

Kircher, Athanasius. 1660. *Iter extaticum coeleste*. Herbipoli: Endter.

Knobloch, Eberhard. 2002. Vielheit der Welten – Extraterrestrische Evidenz. In *Ideale Akademie. Vergangene Zukunft oder konkrete Utopie?* ed. Wilhelm Vosskamp, 165–186. Berlin: Akademie.

Lambert, Ladina B. 2002. *Imagining the unimaginable. The poetics of early modern astronomy*. Amsterdam/New York: Rodopi.

Leeuwenhoek, Anthoni van. 1939–1999. *Alle de brieven*. 15 vols. Amsterdam: Swets and Zeitlinger.

Leibniz, Gottfried Wilhelm, and Samuel Clarke. 1978. Streitschriften zwischen Leibniz und Clarke. In Gottfried Wilhelm Leibniz, *Die philosophischen Schriften*, ed. C.J. Gerhardt, vol. 7. Hildesheim/New Berlin: Georg Olms Verlag.

Leinkauf, Thomas. 2009. *Mundus combinatus: Studien zur Struktur der barocken Universalwissenschaft am Beispiel Athanasius Kirchers SJ (1602–1680)*. Berlin: Akademie Verlag.

———. 2017. *Grundriss: Philosophie des Humanismus und der Renaissance*. 2 vols. Hamburg: Felix Meiner.

Lerner, Michel-Pierre. 2008. *Le monde des sphères*. 2 vols. Paris: Les Belles Lettres.

Locke, John. 1975. *Essay concerning human understanding*. ed. Peter H. Nidditch. Oxford: Clarendon Press.

Mandelbrote, Scott. 2007. The uses of natural theology in seventeenth-century England. *Science in Context* 20: 451–480.

Matytsin, Anthon. 2013. Scepticism and certainty in seventeenth – and eighteenth-century speculations about the plurality of worlds. *Science et Esprit* 65: 359–372.

McKnight, Stephen. 1991. *The modern age and the recovery of ancient wisdom*. Columbia/London: University of Missouri Press.

Mehl, Édouard. 2013. La physique cartésienne et l'interpretation de la Genèse. In *Lire et interpréter. Les religions et leurs rapports aux textes fondateurs*, ed. Anne-Laure Zwilling, 135–149. Genève: Labor et Fides.

———. 2021. Kepler métaphysicien. *Revue des Questions Scientifiques* 192: 261–279.

Meli, Domenico B. 2019. *Mechanism. A visual, lexical, and conceptual history*. Pittsburgh: The University of Pittsburgh Press.

Mémoires de l'Académie royale des sciences, depuis 1666 jusqu'à 1699. 11 vols, 1729–1734. Paris: Compagnie des Libraires.

Methuen, Charlotte. 1998. *Kepler's Tübingen. Stimulus to a Theological Mathematics*. Aldershot: Ashgate.

Mormino, Gianfranco. 2000. Ammirare e comprendere: la concezione del sapere di Christiaan Huygens. In *Potentia Dei*, ed. Gianni Canzani, Miguel A. Granada, and Yves Ch. Zarka, 495–511. Milano: Angeli.

Mormino, Gianfrancesco. 2003. Le rôle de Dieu dans l'oeuvre scientifique et philosophique de Christiaan Huygens. *Revue d'histoire des sciences* 56: 113–133.

Mullet, Isabelle. 2011. *Fontenelle ou la machine perspectiviste*. Paris: Honoré Champion.

Nadler, Steven. 2000. Doctrines of explanation in late scholasticism and in the mechanical philosophy. In *The Cambridge history of seventeenth-century philosophy*, ed. Daniel Garber and Michael Ayers, vol. 1, 513–552. Cambridge: Cambridge University Press.
Newton, Isaac. 1726. *Philosophiae naturalis principia mathematica*. Londini: Innys.
———. 1731. *A treatise of the system of the world*. London: Fayram.
———. 1961. *The correspondence of Isaac Newton*, ed. Herbert W. Turnbull, vol. 3. Cambridge: Cambridge University Press.
———. 2004. *Philosophical writings*, ed. Andrew Janiak. Cambridge: Cambridge University Press.
Osler, Margaret. 1994. *Divine will and the mechanical philosophy. Gassendi and Descartes on contingency and necessity in the created world*. Cambridge: Cambridge University Press.
Osler, Margaret J. 2001. Whose ends? Teleology in early modern philosophy. *Osiris* 16: 151–168.
Palmerino, Carla R. 2006. The mathematical character of Galilei's book of nature. In *The Book of nature in early modern and modern history*, ed. Klaas van Berkel and Arie J. Vanderjagt, 27–44. Leuven: Peeters.
Paracelsus. 1924–1933. *Sämtliche Werke*, ed. Karl Sudhoff, 14 vols. Munich/Berlin: Oldenbourg.
Paracelsus, and Oswald Croll. 1657. *Philosophy reformed & improved in four profound tractates*, translated by H. Pinnell. London: Lodowick.
Pascal, Blaise. 1998. *Œuvres complètes*, ed. Michel Le Guern, 2 vols. Paris: Gallimard.
Pico della Mirandola, Giovanni. 2019. Oration. In Brian Copenhaver, *Magic and the dignity of man. Pico della Mirandola and his oration in modern memory*, 459–482. Cambridge, MA/London: The Belknap Press of Harvard University Press.
Plethon, Georgios G. 1858. *Traité des lois*. ed. Charles Alexandre, trad. Augustin Pellissier, Paris: Didot.
Porta, Giambattista della. 1597. *Magia naturalis*. Francofurti: Wechius.
Poulouin, Claudine. 2013. Introduction. In Bernard de Fontenelle, *Œuvres complètes. Vol. 1: Entretiens sur la pluralité des mondes*, ed. Claire Cazanave and Claudine Poulouin, 59–134. Paris: Honoré Champion.
Ptolemy, Claudius. 1967. The Arabic Version of Ptolemy's Planetary Hypotheses, ed. Bernard R. Goldstein. *Transactions of the American Philosophical Society* 57: 3–55.
Radelet de Grave, Patricia. 2003. L'universe selon Huygens, le connu et imaginé. *Revue d'histoire des sciences* 56: 79–112.
Ray, John. 1735. *The wisdom of God*. London: Innys/Manby.
Régis, Pierre-Sylvain. 1690. *Système de philosophie, contenant la logique, la métaphysique, la physique et la morale*. Paris: Anisson.
Rheita, Schyrlaeus. 1645. *Oculus Enoch et Eliae sive Radius sidereomysticus*. 2 vols. Antverpiae: Verdussius.
Riccioli, Giambattista. 1651. *Almagestum novum*. Bononiae: Benatius.
———. 1665. *Astronomia reformata*. Bononiae: Benatius.
Rohault, Jacques. 1671. *Traité de physique*. Paris: Thierry.
Rossi, Paolo. 1972. Nobility of man and plurality of worlds. In *Science, medicine and society in the Renaissance. Essays to honor Walter Pagel*, ed. Allen G. Debus, vol. 2, 131–162. London: Science History Publications.
Roux, Sophie. 1998. Le scepticisme et les hypothèses de la physique. *Revue de synthèse* 119: 211–255.
———. 2017. From the mechanical philosophy to early modern mechanisms. In *The Routledge handbook of mechanisms and mechanical philosophy*, ed. Stuart Glennan and Phyllis Illari, 26–45. London: Routledge.
———. 2022. What to do with the mechanical philosophy. In *The Cambridge history of the scientific revolution*, ed. David M. Miller and Dana Jalobeanu, 75–95. Cambridge: Cambridge University Press.
Rüfner, Vincenz. 1955. Homo secundus deus. Eine geistesgeschichtliche Studie zum menschlichen Schöpfertum. *Philosophisches Jahrbuch* 63: 248–291.

Sakamoto, Kuni. 2016. *Julius Caesar Scaliger, Renaissance reformer of Aristotelianism: A study of his exotericae exercitationes*. Brill: Leiden.
Sartorius von Waltershausen, Bodo. 1935. *Paracelsus. Am Eingang der Deutschen Bildungsgeschichte*. Leipzig: Meiner.
Scaliger, Justus. 1558. *Exotericarum Exercitationum liber quintus decimus*. Lutetiae: Vascosanius.
Schmaltz, Tad. 2008. *Descartes on causation*. Oxford: Oxford University Press.
Schmid, Stephan. 2010. *Eine Untersuchung der Transformation teleologischer Erklärungen*. Berlin/New York: De Gruyter.
Schmidt-Biggemann, Wilhelm. 2005. Kosmos und Kabbalah: Robert Fludd Naturkonzeption. In *Der Naturbegriff in der Frühen Neuzeit. Semantische Perspektiven zwischen 1500–1700*, ed. Thomas Leinkauf, 212–235. Tübingen: Niemayer.
Schoot, Jan van der. 2014. Interpreting the Cosmotheoros (1698): A historiographical essay on theology and philosophy in the work of Christiaan Huygens. *De Zeventiende Eeuw* 30: 20–39.
Seidengart, Jean. 2006. *Dieu, l'univers et la sphére infinie. Penser l'infinité cosmique à l'aubé de la science classique*. Paris: Albin Michel.
Shanahan, Timothy. 1988. God and nature in the thought of Robert Boyle. *Journal of the History of Philosophy* 26: 547–569.
———. 1994. Teleological reasoning in Boyle's *Disquisition about final causes*. In *Robert Boyle reconsidered*, ed. Michael Hunter, 176–192. Cambridge: Cambridge University Press.
Sheehan, Jonathan. 2003. From Philology to fossils: The Biblical Encyclopedia in early modern Europe. *Journal of the History of Ideas* 64: 41–60.
Siebert, Harald. 2005. The early search for stellar parallax: Galileo, Castelli, and Ramponi. *Journal for the History of Astronomy* 36: 251–271.
———. 2006. *Die grosse kosmologische Kontroverse: Rekonstruktionversuche anhand des Itinerarium exstaticum von Athanasius Kircher, SJ (1602–1680)*. Stuttgart: Franz Steiner Verlag.
Snobelen, Stephen. 2001. God of gods, lord of lords. The theology of Isaac Newton's general scholium to Principia. *Osiris* 16: 169–208.
Stoffel, Jean-François. 1998. La révolution copernicienne et la place de l'homme dans l'Universe. Étude programmatique. *Revue philosophique de Louvain* 96: 7–50.
———. 2001. Géocentrisme, héliocentrisme, anthropocentrisme: quelles interactions? *Scientiarum Historia* 27: 77–92.
———. 2002. La révolution copernicienne responsable du désenchantement du monde? L'exemple des analogie solaires. *Revue belge de philologie et d'histoire* 80: 1189–1224.
Tyard, Pontus de. 1557. *L'univers ou discours des parties, et de la nature du monde*. Lion: Tournes.
Van Helden, Albert van. 1985. *Measuring the universe. Cosmic dimensions from Aristarchus to Halley*. Chicago/London: University of Chicago Press.
Walsh, Kirsten. 2019. Newton's scaffolding. The instrumental roles of his optical hypotheses. In *Experiment, speculation and religion in early modern philosophy*, ed. Alberto Vanzo and Peter R. Anstey, 125–157. New York: Routledge.
Westfall, Richard. 1981. *Never at rest. A biography of Isaac Newton*. Cambridge: Cambridge University Press.
Wilkins, John. 1802a. A discourse concerning a new planet. In John Wilkins, *The mathematical and philosophical works*, vol. 1, 132–261. London: Vernon and Hood.
———. 1802b. The discovery of a new world. In John Wilkins, *The mathematical and philosophical works*, ed. John Wilkins, vol. 1, 1–130. London: Vernon and Hood.
Zambelli, Paola. 2007. *White magic, black magic in the European Renaissance. From Ficino, Pico, Della Porta to Trithemius, Agrippa, Bruno*. Leiden/Boston: Brill.

Chapter 3
The End of the *Contemplator Caeli*

Contents

Abstract Many Greek and Roman philosophers cherished the ideal of the *contemplator coeli*, according to which human beings were meant to observe the sky and nature as a whole. In this ideal, the human eye and nature were teleologically optimized for each other so that humans could see all that was allowed them by gods or God. Hans Blumenberg labelled this openness of nature to human sight as the 'postulate of visibility'. When early modern natural philosophers and astronomers started to use telescopes and microscopes for the investigation of nature, the ideal of the *contemplator coeli* vanished. Both types of instruments invalidated the notion that human senses naturally captured the constitutive elements of nature. Looking through telescopes and microscopes had serious epistemological consequences: It became clear that nature contained an inexhaustible number of invisible objects. Philosophers also realized that humans saw only the surface of reality, behind which lay the immeasurable dimensions of other layers of reality in the microworld and the macroworld. Telescopes and microscopes thus demonstrated that human knowledge had a tiny range and depth relative to nature. By the end of the seventeenth century, however, the limits of human knowledge were no longer perceived as an epistemological fatality but as a horizon for further expansion and development.

Keywords Telescopes · Microscopes · Postulate of visibility · Contingency of nature · Contingency of human experience · Early modern epistemology

D. Špelda, *The Origins of the Idea of Scientific Progress*, International Archives of the History of Ideas Archives internationales d'histoire des idées 250,
https://doi.org/10.1007/978-3-031-60526-0_3

Cosmological anthropocentrism placed humans at the centre of the cosmos. According to this philosophical tradition, however, the centre was not the most important place in the cosmos, because it was the place furthest from the periphery, beyond which dwelt the pagan gods or the Christian God with his angels. The centre was considered the place ontologically and cosmologically most miserable, i.e. most distant from God and most contaminated with matter.[1] The excellence of the centre laid in its anthropological and epistemological function. The centre represented the optically privileged place from which the best view over the cosmos was obtained. The central position enabled human beings to capture the world in its totality, or at least to the maximum extent permitted to humans by God. Renaissance thinkers often described humans contemplating the world from a privileged location by using the Greek metaphor of an observer of the heavens, a stargazer.

In this chapter I first introduce the classical philosophical ideal according to which there was an affinity between human senses and reality. I then explain how the introduction of optical instruments in the early modern period demonstrated the contingency of human experience. Human curiosity was freed from the narrow boundaries defined by the teleology of nature and could gradually begin to satisfy itself in vast regions of the universe and the microscopic world. It quickly became apparent to natural philosophers that knowledge of the immense natural world would take a long time. Therefore, the revealed contingency of human experience may be also regarded as a condition of the possibility of scientific progress.

3.1 *Contemplator caeli*

The Latin term *contemplator caeli* or *spectator caeli* originated in Latin authors who followed the older motifs of Greek philosophy, especially stoicism. The stargazer was an ethical, epistemological, and cosmological definition of humans that connected them teleologically with the world and with gods or God. According to this belief, humans were born to observe the sky. According to Cicero (1967, 257–259), the Providence of nature

> has raised them [human beings] from the ground to stand tall and upright, so that they might be able to behold the sky and so gain a knowledge of gods. For men are sprung from the earth not as its inhabitants and denizens, but to be as it were the spectators of things supernal and heavenly, in the contemplation whereof no other species of animals participates.

The visible heavens embodied the divine and rational order of the cosmos. Exploring heavenly phenomena led humans to the knowledge of the world, God, and ultimately themselves. Thanks to this knowledge, human beings could attain beatitude. Observing the heavens was not meant to be an everyday practical observational activity leading to astronomical discoveries. *Contemplatio caeli* was a fascination with the

[1] The center of the world was the least valuable place in the world. See Brague 1994; Stoffel 2001, 2002.

splendour and order of the sky that brought about individual spiritual perfection. It was not science in the modern sense, but therapy for the soul, which, through the observation of the cosmos and the sublimity of nature, detached itself from the earthly world and drew closer to its heavenly origin.

The idea of *contemplator caeli* had a long history in following periods.[2] It appeared in the Middle Ages and we can find it often in the Renaissance sources. I want to outline several features of the idea that are relevant to the topic of this book.

The first characteristics of the *contemplator caeli* was his readiness for the observation of the cosmos. From the physiological point of view, stargazer had an upright posture, eyes at the front of the head, and a neck that allowed the face to be turned upwards. Cosmologically, the *contemplator caeli* was adjusted to observe the cosmos because he was placed in the midpoint of the closed world. In his *De otio* Seneca stated that the centre provided the best view of the spherical and closed cosmos (Seneca 2010, 225). Epistemologically, stargazers were adapted to the world in the sense that nature gave them an inquisitive spirit (*curiosum nobis natura ingenium dedit*), as Seneca wrote also in *De otio* (Seneca 2010, 222). That is why humans naturally desired knowledge. The visible world, and especially the heavens, by its beauty and splendour, aroused in humans a yearning for knowledge and exploration owing to which they could surpass mere animal survival and come to an understanding of the order and origin of the world. The epistemological teleology of the universe included also the belief that humans could reasonably expect their desire for knowledge to be satisfied.

We can find these anthropological, cosmological, epistemological characteristics of the stargazer in various forms in Renaissance authors such as Bovelles, Cardano, Croll, Giulio Cesare Vanini, Johannes Jessenius, or Comenius.[3] Pontus de Tyard described the ideal in words that were reminiscent of Cicero and Seneca (Tyard 1557, 5):

> For we would be uselessly placed by Nature in the midst of this admirable worldly Theatre, with our face erect, our eyes high and clear, our head flexible on all sides, and our whole person agile in turning round, if we did not look in diligent consideration at all the substances of this World, curved to offer itself more conveniently to our view.

The second typical feature of the idea of a stargazer was its connection with natural theology. A closed and geocentric cosmos appeared completely to the view of the stargazer, who realized that such an admirable order could only be the work of the wise Creator. The natural world was a source of admiration and wonder that elevated humans to the level of the gods and to the imitation of their blissful life (Cicero 1967, 271).

Although *contemplatio caeli* was not intended as a practice of observational astronomy, astronomers readily adopted this philosophical ideal. In classical

---

[2]Probst 2021; Blumenberg 1966; for further context on the history of this idea, see Pfeiffer 2001; Brague 2005, 155–196.

[3]Bovillus 1510, 132v; Cardano 1580, *epistula nuncupatoria*, unpaginated; Vanini 1615, 24–25; Croll 1620a, 58; Jessenius 1601, fol. 11r-v; Comenius 1964, vol. 1, 357b.

antiquity, astronomy was often considered a technical skill rather than true knowledge. By using the idea of a stargazer, astronomers could emphasize the dignity of their discipline: Astronomy brought humans moral perfection, piety, and beatitude since there was no higher subject with which humans could be concerned. According to Ptolemy, his discipline raised humans to the gods and changed their nature 'to a similar spiritual state'.[4]

Renaissance humanism was familiar with the idea of the stargazer, and humanistically trained astronomers used it regularly. In his speech *Oratio de disciplinis mathematicis* (1574), Tycho Brahe took advantage of this classic *topos* to defend astronomy from charges of futility[5]:

> Consider nothing more fitting to the nature of man and more in harmony with the purpose for which man was created and placed on Earth, the centre of the universe, than that, looking out from his central position [...] he may pass his life happily and [...] recognising God as the creator in these his most wise and varied works, should honour, praise, and reverence him deservedly.

Similar praise of astronomy as a way of life that brought humans to beatitude and to God can be found in many other Renaissance and early modern astronomers such as Johannes Regiomontanus, Caspar Peucer, Simon Grynaeus, Georg Joachim Rheticus, Johannes Stadius, Martinus Hortensius, and Michael Mästlin.[6] *Topos* was also found in the writings of Catholic astronomers such as the Jesuits Christoph Scheiner, Christopher Clavius, and Riccioli.[7] Johannes Kepler was even able to transfer the idea of the stargazer into heliocentric cosmology. God wisely granted the Earth a trajectory that was between the trajectories of other planets in order to allow humans 'contemplate all the planets' (Kepler 1938–2009, vol. 4, 309). The introduction of heliocentrism therefore by no means resulted in the rejection of the idea of the stargazer.

The third typical feature of the stargazer was the relationship of reciprocity between the observer and the observed. Humans needed the world to achieve an earthly form of the beatitude of the gods, and the world needed humans for its own completion. Beauty and order could not be spoken of unless there was someone to perceive them. The world had to have an observer; otherwise, all the splendour of God's work would be in vain. In his *De otio* (IV,2) Seneca asked: 'What service to god does the contemplator of such matters provide? That his great works are not without witness' (Seneca 2010, 225). The world was thus incomplete without an understanding observer whose mind would comprehend and admire God's work. The stargazer was a witness who, by his examination of nature, testified to God's goodness, wisdom, and omnipotence.

---

[4]Ptolemy 1998, 37; cf. the commentary in Taub 1993, 34–36.

[5]Brahe 1913–1929, vol. 1, 152; English translation in Brahe 1929, 319–320.

[6]Regiomontanus 1544, fol. 4r; Reinhold 1551, fol. α3; Grynaeus 1533, fol. a2r-3r; Rheticus 1542, fol. A8r-v; Stadius 1560, 18; Hortensius 2006, 103; Mästlin 1578, unpaginated dedicatio; Mästlin 1588, fol. *3. Cf. Kusukawa 1995, 127–142; Roelants 2012.

[7]Clavius 1606, fol. A 3; Scheiner 1630, 593; Riccioli 1651, vol. 1, 1.

The principle of witness is also easily found in Renaissance thought as an important part of religious and philosophical anthropology.[8] Giovanni Pico della Mirandola argued that human dignity laid above all in the ability of humans to provide an understanding reflection on God's work. Man was created as an epilogue of creation 'to assess the reason for so great an undertaking [i.e. the creation of the universe], to love its beauty, to be astonished by its immensity' (Pico della Mirandola 2019, 459). Similar reflections on God's need for an observer after finishing creation were later formulated by Bovelles, Paracelsus, Croll, or Comenius.[9] Kepler also pointed out that by understanding the geometric structure of the cosmos, humans completed the process that began with creation.[10]

Renaissance natural history also held a variant of the classical ideal of the stargazer. Natural historians needed to turn the observer's eyes towards earthly nature and therefore replaced the upward gaze with a view of the theatre of the earthly world, which was prepared for the spectator who came as the epilogue of creation. The metaphor of the theatre of nature prepared for humans appeared in the works of Conrad Gessner, Guillaume Rondelet, Vanini, Cardano, or Bodin.[11]

Discussing the functions and meanings of the theatre metaphor in Renaissance natural philosophy, Ann Blair highlighted three key functions of the metaphor (Blair 1997, 156). Firstly, the idea of the theatre of the world functioned as an argument from design. Secondly, the idea attributed an important function to humans: nature as a theatre required a spectator who bore witness to God's wisdom and goodness. The last function was epistemological. According to Blair, the theatre metaphor 'emphasized especially the vastness and grandeur of nature, its varied and complex harmonies laid out all at once for human contemplation [...]; this theater offers a complete and coherent view of the world in one gaze'.[12]

Following Blair, it can be said that the idea of the theatre of nature justified theoretical optimism. The earthly world could be observed in its completeness and offered itself to the human eye because that was what it was created and prepared for. This meant that neither the idea of the stargazer nor the theatre metaphor included time as a factor of knowledge. Nature itself opened up to humans at once and in its entirety. There were no empirical domains that opened up only over time because nature had no history. The image of the vault of heaven above the human head and the spectacle in the theatre of nature expressed the synchronous presence of everything knowable. In a single glance, humans comprehended the diversity of nature and recognized the existence and wisdom of the Creator.

---

[8] See many examples in Groh 2003, 24–420; Blumenberg 1996b, 309–376.

[9] Bovillus 1510, 51; Paracelsus 1924–1933, vol. I/12, 57; Croll 1620, 56; Comenius 1964, vol. 1, 110a, 299a.

[10] Kepler, letter to Heydon, summer 1605, in Kepler 1938–2009, vol. 15, 235.

[11] Gessner 1551–1555, vol. 1, fol. α3v, α4v; Rondelet 1558, fol. a3; Cardano 1580, *Epistula nuncupatoria*, unpaginated; Vanini 1615, 24; Bodin 1597, 633: [Deus] *hominem non in abdito angulo, sed in mundi medio collocavit, ut multo facilius ac melius quam in coelo rerum omnium universitatem contemplaretur,& eius opificia omnia...*

[12] Blair 1997, 154–155, see also Leinkauf 2009, 31, 110.

Certainly, some things were invisible and inaccessible, but their invisibility resulted from the decision of God, who did not want humans to know certain things. The cognitive optimization of the world for humans thus manifested itself also in the setting of the limits of human knowledge. The effort to penetrate further embodied curiosity that was futile and prideful because it wanted to penetrate fields the knowledge of which God had reserved for himself (Bodin 1597, 609). Even though the scope of human knowledge was limited, natural philosophy had neither the ambition nor the ability to push its knowledge beyond the visible, because it relied on the optimizing work of God's Providence who provided for humans to know everything they needed to know.

The fourth feature of the idea of stargazer was a teleological relationship between the human senses and the natural world. The human body, including the sensory organs, was set up in such a way that humans could perceive everything that God considered appropriate. Blumenberg coined the concept of the postulate of visibility (*Sichtbarkeitspostulat*) for this belief (Blumenberg 2002, 15; Blumenberg 1986, 113–114). In his view, this assumption was originally part of the broader anthropocentric philosophical framework of Greek philosophy according to which nature fully offered itself to the human senses at every moment (providing that its extent remained the same throughout eternity). The Greeks, according to Blumenberg, believed that they were spectators of the world and that humans had an optically privileged position in the order of nature. Hence the Greek philosopher was far from suspecting 'that his eyes and the world could be so disproportioned to each other that essential regions and objects, in reality, would be withheld from man's access.'[13]

Blumenberg's general remark can be substantiated by conclusions made by historians of theories of perception, vision in particular. Ancient and medieval conceptions of perception can mostly be characterized as perceptual optimism or realism. According to this view, human senses represented the world as it was and to its full extent unless God or gods decided that humans had no right to perceive a specific piece of reality.

In the history of theories of vision, there have been two basic approaches—extramission and intromission. The extramission theory assumed that rays radiated from the eye, while the intromission theory assumed that entities enabling perception entered the eye. In classical antiquity, the theory of extramission was developed specially by mathematicians who were interested in the geometrical properties of rays and wanted to explain visual phenomena with the help of geometry in the framework of instrumentalist methodology 'to save the phenomena' (Smith 2018, 47–53). By contrast, the theory of intromission focused on the philosophical and physiological aspects of perception. It was characterized by perceptual realism according to which humans perceived the properties that objects actually have (Hamou 2002, 10, 27–28).

[13]Blumenberg 1996a, 726, English translation in Blumenberg 1987, 625.

The most influential intromission theory was provided by Aristotle (Caston 2005). Aristotle understood perception as a kind of qualitative change induced in the perceiver by the perceived object through sensible forms that enter the sensory organ.[14] After receiving the sensible forms, the perceiving organ was likened to an external object. Aristotle understood this process as an actualization of potentiality which took place in both the perceived object and the perceiving organ. When sensible forms were not perceived, they were only the potentiality of the object to act on the corresponding sense. Sense, in turn, was only the potentiality of the perceiving being to let the sensible forms act upon itself. Thus, perception was achieved only when these two potentialities meet.[15]

Objects had different properties that related to different senses. Aristotle called these properties 'special sensibles'. Examples were colours that were perceptible only to the sense of vision and sounds accessible only to the sense of hearing. Aristotle further distinguished 'common sensibles' that were perceptible to multiple senses, such as motion and rest. The third kind of properties was 'accidental sensibles', for example when perceiving that an observed white object of a certain size was a person. From the epistemological point of view, the most important thing about this distinction is that when 'special sensibles' were perceived by their corresponding senses, the perception was veridical.[16] Aristotle did not claim that our perceptions were always correct. Perception was veridical when the sense perceived its 'special sensibles'. For example, the eye was never mistaken about colour.[17]

I would like to briefly outline a few aspects of the Aristotelian theory of perception that were related to the belief that the constitutive elements of nature were accessible to the human senses—and thus that human perception was not contingent.

The first aspect was Aristotle's idea that there were no more than five senses.[18] Their number and arrangement were sufficient to enable humans to grasp the variety and essence of material nature. Pavel Gregorić explained that in Aristotle's conception we have five senses in order to be able to perceive five different types of special sensibles: colours, sounds, odours, flavours, and tangible qualities. According to Gregorić, 'the five different types of special sensible, for which we have five different senses, are *all* such properties that exist in the universe'.[19] There was no sixth type of special sensible designated for a sixth sense. The quintet of senses exhausted all the properties of the world. Reality contained nothing perceptible for

[14] Aristotle, *De anima* II,5, 416b33-4, English translation in Aristotle 1995, vol. 1, 663.

[15] Aristotle, *De sensu* 438b12-16, English translation in Aristotle 1995, vol. 1, 696.

[16] Aristotle, *De anima* II,6, 418a8-16, English translation in Aristotle 1995, vol. 1, 665; cf. III, 3, 428a8-18, English translation in Aristotle 1995, vol. 1, 681.

[17] Hamou 2002, 17; cf. Lindberg and Steineck 1972.

[18] Aristotle, *De anima* III, 1, 424b20-21, English translation in Aristotle 1995, vol. 1, 675.

[19] Gregorić 2019; cf. Hamou 1999, vol. 2, 106–108.

which there was no corresponding sense and sensory organ. Voltaire's character Micromégas with his thousand senses was unthinkable in the Aristotelian world.

Secondly, according to Aristotle's conception of perception, the human sensory organs could give an adequate picture of reality, because they were potentially 'sensible forms' that arose from things themselves. There was proportion and readiness between the knower and the known. Hence the Aristotelian philosophy could barely conceive of the perceptibility of an invisible microworld that was different from the visible macroscopic world. Aristotle's theory of sense perception was therefore incompatible with atomist ontology.

In a passage in *De generatione et corruptione*,[20] where Aristotle tried to refute atomism, he dealt with mixing various materials. Christoph Lüthy (1996, 11–12) explained that according to Aristotle, matter appeared to the senses in the form of pairs of qualities carried by the four elements. When elements were mixed to form a more complex object, they lost their individual qualities and acquired a new set of qualities. According to Aristotle, such a compound was homogeneous: 'the compound must be uniform—any part of such a compound being the same as the whole, just as any part of water is water.'[21] Nothing could be expected from a visual examination of the invisible small parts of homogenous compounds because they could not reveal anything else than what the naturally perceptible parts showed. The microscopic particles of water were still the same water that was visible by natural sight. The basic constituents of the material world were therefore indifferent to magnification in Aristotelian thought. There was nothing particular to see beyond the limit of natural visibility.

Aristotle speculated further that if atomist ontology was valid, atoms of two starting materials would remain intact within a new compound but humans would see this heap as a homogeneous entity. Aristotle continues with a contrafactual reflection: Let us imagine an observer with abnormally penetrating vision. Such an observer would see that a mixture was made up of two groups of atoms from two starting materials, while ordinary humans would see a homogenous compound.[22]

In Aristotle's view, if atomism was true different compounds would be relative to the sharpness of vision. Some observers would see a homogenous compound, and others endowed with much sharper eyesight would spot heaps formed of two kinds of atoms. Humans and other beings thus would perceive the same object differently depending on how their senses were arranged. This was exactly the conclusion that early modern philosophy came to after telescopes and microscopes began to be used. In contrast, Aristotle wanted to preserve the universality and reliability of human sensory experience. His ontology insisted that true compounds were homogeneous and would therefore appear the same to all observers regardless of their visual acuity.

---

[20] Aristotle, *De generatione et corruptione*, 328a6-328a18, English translation in Aristotle 1995, vol. 1, 536.

[21] Aristotle, *De generatione et corruptione*, 328a11-12, English translation in Aristotle 1995, vol. 1, 536.

[22] Aristotle, *De generatione et corruptione*, Ibid. 328a14-16, English translation in Aristotle 1995, vol. 1, 536–537.

Enhancing vision with instruments could not change this. As Lüthy aptly noted, 'the Philosopher had [...] banned microscopy from philosophical discourse even before it had become a practical option'.[23]

The third aspect of Aristotle's perception theory concerns the observation of the stars. The postulate of visibility was also valid for the observation of distant objects. It was consistent with the Greek belief in a finite cosmos in which the range of human senses corresponded to the limits of the world. The sensory images of the fixed stars represented the edge of a finite and closed world, which was constituted by the sphere into which they were embedded. Blumenberg explained[24]:

> The postulate of visibility follows from the symmetrical construction of the geocentric universe and man's central position in it. If all the points of the heaven of the fixed stars are equally distant from the Earth, the suspicion has to be excluded that there could be stars that, on account of their distance, remain invisible for men: If any fixed star at all can be perceived, then all of them can be. There is no conceivable need for an auxiliary means that could have promised to accomplish more here than the human eye can accomplish.

Thus, similarly to Lüthy, Blumenberg indicated that neither the telescope nor the microscope was conceivable by Aristotelian philosophy because there would be nothing to observe in these instruments.

In the Middle Ages, the philosophical intromission theory was combined with the mathematical extramission theory, which dealt with the geometry of visual rays. Justin L. Smith summarized his interpretation of medieval perspectivist theory by asserting that

> the theory resulting from this fusion presupposes that physical reality and our mental 'picture' of it are in strict conformance with each other. In short, objective reality actually is as we visualize it in our minds [...] this presupposition was not new with perspectivists. From at least the time of Plato and Aristotle, every effort, whether philosophical or mathematical, to explain how we apprehend the physical world by means of vision was predicated on it (Smith 2018, 276).

I think that it is possible to approximately identify Blumenbergian *Sichtbarkeitspostulat* with the paradigmatic 'presupposition' mentioned by Smith. Then we can also argue that a significant part of the premodern philosophical tradition held that human perceptual faculties were adapted to grasp nature in its constitutive elements. In such a context, however, it was difficult to think of moving the limits of empirical knowledge. What humans could see and know was optimized for them. The sensory experience was not contingent, and its extent did not depend on human effort. The extent of knowledge was rigidly and statically limited by the boundaries of what God or nature chose to make available to human senses. After the invention of the telescope, this teleology ceased to be considered convincing.

---

[23] Lüthy 1996, 12; Hamou 1999, vol. 2, 106–107.

[24] Blumenberg 1996a, 731, English translation in Blumenberg 1987, 629. cf. Hamou 1999, vol. 1, 133.

3.2 Contingency of Human Experience

In the second chapter, I tried to show how early modern mechanistic natural philosophy brought about the contingency of nature. In this section, I want to demonstrate how early modern philosophy rejected the postulate of visibility and insisted on the contingency of human perception.

Descartes's theory of perception was based on his ontology and physiology, and therefore he explained the process of perception mechanistically. The light that the human eye saw was actually impulses of rapidly rotating particles that formed images on the retina. Animal spirits transferred these retinal paintings through hollow nerve fibres to the inner folds of the brain in the form of various configurations of movements that could be likened to vibrations. These vibrations were structurally isomorphic to the pictures of things on the retina and were transferred to the pineal gland, which transmitted them to the mind.[25]

God set up the relationship between the senses, the brain, and the mind in this way, but this did not mean that human vision depicted the real properties of the external world. Descartes rejected the idea that the configurations of vibrations that were transmitted to the folds of the brain could be considered an image resembling the external world. Humans did not know whether the sensory image of reality in their minds was fictional. The comparison with the blind man's cane at the beginning of *La Dioptrique* (1637) illustrated that humans did not feel or perceive external objects directly; impulses were felt, much in the way a blind man felt the trembling of his cane.[26]

> You have only to consider that the differences a blind man notes between trees, rock, water and similar things by means of his stick do not seem any less to him than the differences between red, yellow, green and all the other colours seem to us. Any yet in all those bodies the differences are nothing other than the various ways of moving the stick or resisting its movements. Hence you will have reason to conclude that there is no need to suppose that [. . .] there is something in the objects which resembles the ideas or sensations that we have of them. . .

As Smith remarked, these 'ideas or sensations' were nothing like the impulses and spins of corpuscules that gave rise to them (Smith 2018, 410). Therefore many qualities of the human perceptual image of the world do not exist in this world. For instance, colours were just ideas in the human mind corresponding to the speed at which particles located in a given line of light impulse rotated. The spectrum of colours, therefore, corresponded to the spin of the particles; for instance, the colour red corresponded to the fastest rotation (Smith 2018, 410).

In Descartes view, visual perception expressed the state of the body rather than the veridical image of the external world. Descartes pointed out that when, for example, fumes affected the brain, people might hallucinate. Similarly, damage to

---

[25] For more details see Wolf-Devine 2000; Hamou 2002, 51–53; Hamou 1999, vol. 1, 271–288; Gaukroger 2002, 197–213; Ben-Yami 2015, 11–74.

[26] Descartes 1996, vol. 6, 85, English translation in Descartes 1984–1991, 153.

nerve fibres affected the outcome of visual perception (Descartes 1996, vol. 6, 141). Therefore, Hamou could rightly point out that in Cartesian epistemology, sensory perception was affected by somatic contingency (Hamou 2002, 43–44, Hamou 1999, vol. 1, 271–276). That meant that the sensory image of the world was the unique result of the current physiological disposition of an individual body. If the parts of the eye, i.e. the vitreous, the lens, or the cornea, were arranged and adjusted differently, we would certainly get a different picture of external things. In Descartes's view, different humans (and other beings) perceived the world differently according to the various physiological arrangement of their organs (Descartes 1996, vol. 6, 147–165). This was precisely the conclusion that Aristotle wanted to avoid by hoping for the homogeneity of matter and the universality of human experience.

Descartes's theory of vision presented human vision as contingent—and therefore also as changeable and perfectible. When individual humans perceived the world with slight variations according to the settings of their organs, their vision could be improved by optical devices adapted to their physiological condition and the structure of their eyes. In *La Dioptrique*, Descartes argued that after all, the goal of dioptrics was to correct defects by using 'artificial organs' (Descartes 1996, vol. 6, 165). Descartes and other early modern thinkers after him thus comprehended the eye as a specific type of optical device with a specific alignment that could be adjusted by means of instruments (Hamou 2002, 42).

The possibility of technical perfecting vision had two implications. The absence of any 'natural' or 'absolute' visual perception of the world denied the existence of fixed confines to human visual experience. At the same time, it implied the possibility of a future expansion of empirical knowledge. This is also why the contingency of human experience was a condition of possibility for the emergence of the idea of scientific progress.

In Descartes's philosophy, another idea originated that was important for the epistemological tradition of the second half of the seventeenth century. In the sixth meditation, Descartes wrote that perception was not meant for the acquisition of theoretical knowledge, but had been set up by God to be sufficient for self-preservation (Descartes 1996, vol. 7, 83, 87, vol. 8, 35–36). As Hamou demonstrated, this opinion was widely shared by early modern philosophers (Hamou 2002, 43–46). Not surprisingly, the idea was repeated by Descartes's followers such as Géraud de Cordemoy, Pierre-Sylvain Régis, and Port-Royal's *La Logique* (1662).[27] According to Nicolas Malebranche, 'the senses were given to us, not to know the truth of things in themselves, but only for the preservation of our body'.[28] Leaving the Cartesian tradition, the same idea appeared in Gassendi, Robert Hooke, John Locke, and George Berkeley, and also in Huygens's account of extraterrestrials.[29]

[27] See Cordemoy 1666, 100; Régis 1690, vol. 1, 126–127, 162–164, 261–262; Arnauld and Nicole 1854, 265–268.

[28] Malebranche 1979, vol. 1, 53; English translation in Malebranche 1997, 24.

[29] Gassendi 1658, vol. 3, 312b; Hooke 1705, 8; Locke 1975, 305; Berkeley 2009, 39; Huygens 1888–1950, vol. 21, 719, 725.

This interpretation of perception implied that humans perceived reality not as it was but in relation to their bodies and their survival. This idea substantially changed the philosophical understanding of the world. In the tradition of anthropology based on the idea of the *contemplator coeli*, humans were meant to see everything in the world. For the Greek philosophers, the world was a beautifully ordered cosmos meant to be the object of human contemplation leading to truth and happiness. In view of early modern philosophers the self-preserving function of the sensory organs worked as a principle of receptive economy, established by God, that did not bother humans with data irrelevant to their self-preservation. For instance, humans did not see mites because these animals had no relevance to their bodily self-preservation.

From the self-preserving function of human perception, further consequences followed. In late seventeenth century natural philosophy, the Cartesian idea spread that the eye was similar in structure to an optical device. For example, Malebranche stated that the eyes were 'natural spectacles' whose powers depended on the composition and shape of the lenses and their distance from the retina. Therefore it was not possible to be sure 'that there are two men in the world who see objects as having exactly the same size'.[30]

The dependence of human vision on the physiological arrangement of individual bodies resulted in the idea that individual humans and other beings had their senses set differently. In his *Observationum microscopicarum centuria* (1656), Borel emphasized that human understanding of size did not need to be absolute, since animals observed through a microscope saw the world in their own way as if they were looking into a microscope. In a Cartesian vein, Borel added that different creatures saw the world differently, according to the arrangement of their organs.[31]

Malebranche also thought that microscopic mites had their own microscopic eyes through which they had to have seen the world differently from humans 'since otherwise they would not receive the impressions necessary for the preservation of their life'.[32] If human eyes were set with the magnification that mites have, they would not perform their self-preservation function because humans would see the external world from a perspective that had no relation to the self-preservation of their own bodies. The idea of 'microscopic eyes' was widely shared by early modern philosophers, as shown by Hamou.[33]

Reflections on the setting of human eyes had important epistemological implications. Malebranche noted that we cannot rely on our eyes, i. e. 'natural spectacles', when we want to know the absolute sizes of bodies because they show the world in magnification that corresponds to the needs of our bodies. Humans and their senses were therefore not the measure of things. That things were small in relation to humans did not mean that they were small from the perspective of other beings.

---

[30] Malebranche 1979, vol. 1, 58, English translation in Malebranche 1997, 28.

[31] Borel 1656, fol. a3v: *haec omnia alia animalia varie juxta organorum varietatem percipiant.*

[32] Malebranche, 1979, vol. 1, 58; Malebranche 1997, 28; similarly Locke 1975, 305–306; Berkeley 2009, 51–52.

[33] Hamou 1999, vol. 1, 277–289; Hamou 2002, 43.

Malebranche pointed out that nothing was either large or small to itself (*rien n'est grand ni petit en soi*). The human 'is not an absolute standard against which one should measure other things'.[34] Arnauld's and Nicole's *Logique* put the problem even more directly: We do not know what the true, absolute, and natural size of things is. Our eyes function like optical instruments (*lunettes*), and we do not know whether they reduce or enlarge the objects we see. Perhaps the instrument that we thought magnified or reduced actually represent things at their true size. 'The absolute and natural size of bodies is not known with certainty.'[35]

Early modern philosophy thus abandoned the anthropocentric idea that human visual perception portrayed the world completely and veridically. In the second half of the seventeenth century, perceptual realism was no longer convincing. Natural philosophers agreed that humans did not have optically privileged access to reality. It turned out that there was no natural worldview—only partial perspectives of species and individuals according to the settings of their sensory organs. This view resulted in scepticism concerning empirical phenomena: By means of their senses, humans were not able to identify causes and substances of natural objects.

3.3 Unknown Extension of Reality

Telescopic discoveries brought an explosion of empirical reality hitherto closed between the sphere of the fixed stars and the perceptible qualities of Aristotelian elements. When describing his illustration of part of the constellation Orion, Galileo wrote that at first he had decided to depict the entire constellation of Orion, 'but overwhelmed by the enormous multitude of stars and a lack of time, I put off this assault until another occasion. For there are more than five hundred new stars around the old (*veteres*) one, spread over a space of one or two degrees'.[36] Anton Schyrleus of Rheita, who 30 years later had a better telescope, wrote that with the aid of his instrument he could count twice as many stars in Orion alone as the Ancients counted in the whole sky (Rheita 1645, 197). In his *De vero telescopii inventore* (1655), Borel summarized that the old astronomers knew only 1022 stars (the number of stars in Ptolemy's catalogue), Galileo increased the number of stars observed to 2000, and later Rheita observed 2400 stars in Orion alone (Borel 1655, 17; cf. also Gassendi 1658, vol. 1, 505a). Riccioli estimated at mid-century that there must be about 6000 stars in Orion, and the total may be about 2,000,000 (Riccioli 1651, vol. 1, 413a). By the end of the century, it was clear that the number of stars was inexhaustible and their visibility depended on the performance of telescopes.

[34] Malebranche 1979, vol. 1, 65; English translation in Malebranche 1997, 31.

[35] Arnauld and Nicole 1854, 268; cf. Hallyn 1992, 116–118.

[36] Galilei 1890–1909, vol. 3, 76; English translation in Galilei 1989, 59.

Danish astronomer Ole Rømer worked in the 1670s at the Paris Observatoire. In the autumn of 1676, he reported to the Académie that light did not spread instantaneously.[37] This view was later shared by Newton. In his *Philosophiae naturalis principia mathematica* (1687), Newton wrote that astronomers had confirmed that light arrived from the Sun to Earth in 7 or 8 minutes (Newton 1725, 225). English astronomer Francis Roberts wrote in 1694 that the nearest star must be six thousand times farther than the diameter of Earth's orbit around the Sun. To better approximate this figure, he stated that the journey of light from the stars took longer than the 6-week journey from the West Indies (Roberst 1694, 103). Clarke, in his notes to the English translation of Jacques Rohault's *Traité de physique* (1723), further developed this idea. Referring to John Flamsteed's observations, Clarke estimated that light travelled to Earth from the stars for at least 40 days, sound for no less than fifty thousand years, and the cannonball for a much longer time (Rohault and Clarke 1735, vol. 2, 66).

The discovery of the speed of light and sound (Fontenelle and Godin 1733, vol. 1, 215) did not just mean a further expansion of the cosmos. The discovery further confirmed the disproportion of human perception to the world. The starlight humans were seeing was weeks, months, or years old. Rømer's discovery cancelled the assumed synchronicity of the universe with human sight. As Blumenberg observed, the sky became a mere appearance, an optical simultaneity of non-simultaneous phenomena, 'an accidental section through the many layered depths of huge temporal differences'. A new dimension of invisibility emerged that no longer depended on the quality of telescopes, but solely on time, as observers had to wait until starlight reached them.[38] The speed of light thus demonstrated the limitations of human cognitive equipment to reality. Human earthly experience was perpetually delayed with regard to the universe.

The postulate of visibility was also abandoned in the microscopic world, where the instruments opened up a disturbing vision of another invisible world. In recent literature, Philippe Hamou and Catherine Wilson have provided the most thorough reflection of the epistemological significance of the early modern microscopy. Both historians agreed that the Aristotelian paradigm assumed the adequacy of sensory organs to the world and the indifference of the material world to magnification. In such a paradigm, the need for magnification had to be invented first. Therefore the microscope was spread more reluctantly and with less glory than the telescope in early modern European culture. This reticence was also caused by the fact that the examination of flies, lice, and mites was quite remote from the sublime observation of the heavens in which God and the angels dwelt.[39] The use of the microscope, then, was not so much related to advances in technical design, but to the emergence of a natural philosophy that could work with the microscope as a tool that had a meaningful function.

[37] Cohen 1940; Fontenelle and Godin 1733, vol. 1, 212–216.

[38] Blumenberg 1996a, 630–631; English translation in Blumenberg 1987, 542.

[39] Hamou 1999, vol. 2, 95–116; Wilson 1995, 215–219.

Mechanism was such a philosophy. The mechanistic natural philosophers believed that microscope could finalize the mechanistic programme in ontology. The microscope became an instrument of expectation because the philosophers hoped to observe the corpuscles that formed the visible world. Through direct observation of particles, mechanistic explanations would cease to have the status of mere speculative hypotheses. The microscope was thus considered a 'corpuscular tool' (Lüthy 1996, 14), which was to provide an 'ocular attestation of corpuscularism' (Hamou 1999, vol. 2, 110–111). Evidence of this hope can be found in abundance in early modern literature. The belief in the microscopic revelation of the fundamental constituents of the material world was shared by Galileo, Descartes, Borel, Boyle, Henry Power, or Christopher Wren.[40] Hooke believed he would in time discover other aids to the eye by which 'we may perhaps be able to discover living Creatures in the Moon, or the other Planets, the figures of compounding Particles of matter' (Hooke 1665, fol. a2v).

This hope was not fulfilled, however, and disappointment came at the end of the seventeenth century (Wilson 1995, 218; Wilson 1988). In a letter from 1679, Antoni van Leeuwenhoek reported that water particles would be 'inconceivable for us. And I believe that no man will ever advance so far in science as to be able to see components of water'.[41] Similarly, Locke realized that man did not have 'microscopical eyes' that allowed him to see corpuscles and was therefore dependent in his perception on the slice of reality offered by his sensory organs.[42]

Although microscopes did not show atoms to the human eye, they gave it a different picture: the inexhaustible wealth of microscopic life forms. Microscopes revealed ever newer regions of the microscopic world and their unexpected inhabitants—for example, mites that carried their own mites on their bodies.[43] In *Micrographia* (1665), Hooke wrote that similarly to telescopes, microscopes revealed 'a new visible World'. Hooke further stated that 'in every little particle of [...] matter we now behold almost as great a variety of Creatures as we were able before to reckon up in the whole Universe it self' (Hooke 1665, fol. a2v). The discovery of new living worlds in drops of water was a source of wonder until the end of the seventeenth century. In his *Les Caractères*, Jean de La Bruyère recommended his readers to observe a single drop of water with a microscope, 'and an almost countless number of animalculae will be perceived, moving about with incredible agility, like so many monsters in the vast ocean' (Bruyère 1885, 487).

It turned out that the microscopical world was not the fundament of reality providing an explanation of the macroworld. Christoffer Basse Eriksen explained

[40] Galilei 1960, 105; Descartes 1996, vol. 6, 226–227; Gassendi 1658, vol. 2, 560a; Boyle 1775, vol. 5: 680; Power 1664, fol. c2v-c3r; Wren 1721, 205; cf. Hamou 2004, 106.

[41] Leeuwenhoek 1939–1999, vol. 3, 54–55; quoted in Basse Eriksen 2018, 172.

[42] Locke 1975, 285, 584; cf. Hamou 1999, vol. 2, 277–284.

[43] Malebranche 1979, vol. 1, 55; Arnauld and Nicole 1854, 269; Rohault 1671, 144–145, Hooke 1665, fol. a1v.

that 'the sub-visible world is understood as another part of nature to be observed, and not one fundamentally underlying the visible world' (Basse Eriksen 2018, 21). The invisible microworld became a vast and autonomous part of nature that could not be reduced to the substantive basis of the macroworld. Instead of reaching the ultimate corpuscular level of reality, the microscope only revealed more and more layers of the invisible (Basse Eriksen 2018, 115–116).

Early modern perception theories and the development of observational technology thus proved that humans did not occupy a privileged position in the cosmos because the limits of the human senses had not coincided with the limits of the creative prolificity of nature. In fact, the reach of senses separated humans from two worlds whose limits they were unable to imagine or comprehend.

In Pascal's *Penseés* (1670), we can find a well-known fragment about the two infinities and the disproportion of man to nature, which was probably inspired by several Gassendi's texts, as René Jasiński showed (1933). In this fragment, natural knowledge was presented as 'a great source of humiliation' to humans. Pascal inferred the humiliation of humans from the fact that the Earth and its orbit around the Sun was 'only a delicate point' compared to the stars in the heavens. Not only the solar system, 'the whole visible world is merely an imperceptible speck in nature's ample bosom' that transcended human imagination and reason.

Leaving the world of stars and planets, Pascal presented 'a new abyss' in the microworld. Mites had their own organs, blood in their veins, humours in their blood, drops in their humours, and vapours within the drops. These 'miniature atoms' hid 'an infinity of universes, of which each has its own firmament, planets, and earths in the same proportion as in the visible world'. These microverses had their mites, which in turn consisted of organs, blood, vapour, and atoms, which contained other worlds. According to Pascal, in relation to these layers of reality, humans meant nothing: 'For in the end, what is humanity in nature? A nothingness compared to the infinite, everything compared to a nothingness, a mid-point between nothing and everything, infinitely far from understanding the extremes; the end of things and their beginning are insuperably hidden for him in an impenetrable secret.'[44]

Pascal used this reflection to articulate his conviction that humans should turn inwards and seek God there. Thus, when he wrote about the opening of 'a new abyss' of the infinite, he was in fact using contemporary empirical research for the purposes of Christian apologetics. However, he was not merely drawing attention to the nothingness of humans. He also wanted to show the cognitive disproportion of humans to the world they wanted to explore. Such an endeavour was fundamentally misguided, according to Pascal, because humans were not equipped with sufficient means to comprehend empirical nature. Humans (especially libertines), according to Pascal, presumptuously set out to investigate nature 'as if we had some proportion with it.' But this was an illusion, all things actually emerged from the double infinity of nature: 'this is why we see that every science is infinite in the scope of research.' A

[44]Pascal 1998, vol. 2, 608–610; English translation in Pascal 1995, 67; cf. Witthaus 2005, 177–195.

human being was 'in the mid-point between two extremes' but lacked 'infinite capacities' adequate to both infinities. 'Our intelligence holds the same rank in the order of intelligible things as does our body in the whole vastness of nature.' This natural state made humans 'incapable of certain knowledge or absolute ignorance'.[45]

Pascal thus articulated one of the forms of scepticism related to investigation of the empirical world. His point, however, was not epistemological. The disproportion between humans and the world served Pascal primarily as an objection to the relevance of science, which was ultimately unable to give humans anything but doubt and despair. In Pascal's Christian apologetics, knowledge of nature was vain curiosity that sought something inadequate to humans: 'For mankind's chief malady is its uneasy curiosity (*curiosité inquiète*) about things it cannot know. And it is not so bad for it to be wrong as so vainly curious.'[46] The infinite world was proportioned only to its Creator. The universe comprising two abysses of infinity was terrifying. Man located 'between two gulfs of the infinite and void, will tremble at nature's wonders. I believe that our curiosity turning to admiration, we will be more disposed to contemplate them in silence than arrogantly search them out'.[47] Humans thus should turn to what was appropriate for them. The view of the infinite universe thus was primarily meant to turn man away from science and back to God (Mullet 2011, 54).

Not surprisingly, Pascal followed Cornelius Jansen in these reflections. For Jansen, who in turn followed his model, Augustine, the noble name of science hid also only restless curiosity (*curiosité inquiète*). The investigation of nature was fundamentally wrong, whatever it was focused on. For Jansen, the curiosity with which a philosopher examined a fly was the same as examining an elephant: 'a symptom of the same disease' (Jansenius 1642, 42–43, 47).

Christian apologists understood the observation of nature through optical instruments as a deeply disturbing experience and as an expression of illegitimate curiosity that distracts one from what is truly important. Malebranche followed this interpretation. In the context of sketching a preformation theory of generation, he also warned against exploring the microworld[48]:

> We are justly apprehensive about wanting to enter too deeply into the works of God. In these works nothing but infinities are found everywhere; and not only are our senses and imagination limited to comprehend them, but even the mind, as pure and detached from matter as it is, is too coarse and feeble to penetrate the smallest of God's works. It loses itself, is distracted and dazzled, and is afraid of the sight of what according to the language of senses is called an atom.

In Malebranche's view, the immense natural world caused two reactions which were both bad: it attracts a vain curiosity, or scares humans. While the philosophical

[45] Pascal 1998, vol. 2, 610–611; English translation in Pascal 1995, 68–69.

[46] Pascal 1998, vol. 2, 798; English translation in Pascal 1995, 136.

[47] Pascal 1998, vol. 2, 609; English translation in Pascal 1995, 67.

[48] Malebranche 1979, vol. 1, 57, English translation in Malebranche 1997, 27; similarly Arnauld and Nicole 1854, 269.

tradition believed in the sensory and intellectual kinship of the human mind with the world, Malebranche's words show that this kinship disappeared as a result of the use of optical instruments.

Let us summarize the epistemological and anthropological impact of optical devices. Hamou noted that what appeared to be the framework of reality had become a mere backdrop that masked the true depth of the world. The world alienated from humans. What seemed to constitute its real fabric (the sensible qualities, the homogeneous Aristotelian elements; the vault of the starry sky), revealed itself to be only a kind of coarse decoration placed hiding humans from the authentic depth of the world:

> There is undoubtedly an intellectual victory in this critical crossing of the appearances: the nebulae mask stars, the dead matter is swarming with unnoticed living beings, the homogeneous is revealed to be discrete, the qualitative is the quantitative [...] but this victory is bitter because one cannot let it rest on anything: there is no stop in the optical resolution of the world... (Hamou 2002, 44–45).

Natural philosophers originally expected optical instruments to penetrate the inner structure of things, but it gradually became clear that humans saw in them only surfaces, phenomena, and representations—as the effects of some hidden mechanism. Renaissance natural philosophy often assumed the validity of the signature theory, according to which the nature and properties of things were known by looking at the surface that bore *signa*. The use of optical devices in natural philosophy caused that confidence in the epistemological importance of the senses and the surface of things lost its credibility.

Wilson aptly wrote that the microscope took away 'the privilege of the surface'. What an object looked like on the surface provided no clue to its nature and operations. The key to understanding the powers of an object lay in its internal invisible structure, which represented a new world that defied human analogies, proportions, and natural language. According to Wilson, this world was discovered by accident, as the result of an invention that allowed humans to see something that was not naturally accessible to them. 'The secrets of things were, in this sense, not meant for us to find out at all; we are always, in this new world, intruders. That nature is a system of signs meant for us to read is no longer a tenable view. Nature is not a book' (Wilson 1995, 62–63).

Blumenberg also argued that for natural philosophy, everything visible eventually became a mere phenomenon that was just a haphazardly discovered manifestation of some inaccessible internal structure. In terms of scale, the human senses captured only a contingent slice of reality between the infinite abysses of the microworld and the macroworld[49]:

> The visible world is not only a tiny section of physical reality but is also, qualitatively, the mere foreground of this reality, its insignificant surface, on which the outcome of processes and forces is only symptomatically displayed. Visibility is itself an eccentric configuration, the accidental convergence of heterogeneous sequences of physical events. [...] Optics has

[49]Blumenberg 1996a, 746; English translation in Blumenberg 1987, 642–643.

> become an aspect of objects that does not make a difference: an appearance of appearances. What is accessible at all in this way has turned out to be an arbitrary section through reality, having little significance in relation to what can only be indirectly inferred from it.

How do these epistemological implications relate to the genesis of the idea of scientific progress? Hamou demonstrated that, in the end, the early modern experience of using optical instruments was ambivalent. On the one hand, the instruments showed a contingency of human sensory experience that did not capture the elements of which the world was actually composed. On the other hand, they showed that human experience was not confined within the boundaries of the natural range of our senses and that there were new worlds open to human exploration. The first position represented scepticism; the second embodied progressivism.[50] I think that both positions were closely related: the idea of scientific progress was means of overcoming passive scepticism.

Scepticisms about knowledge of the causes and substances of things was widespread in natural philosophy at the end of the seventeenth century.[51] Scepticism was a humble position that recognized the contingency of nature and pointed out the impenetrability of the majestic work of God to limited human reason. At the same time, scepticism left open the possibility for the continual perfectioning of imperfect human knowledge. The idea of scientific progress thus allowed natural philosophy to move beyond mere scepticism. Natural philosophers could understand the inadequacy of the empirical knowledge as a circumstance that would change over time. Its pettiness was no longer perceived as an epistemological fatality but as an occasion for further expansion and development. The new temporal regime of knowledge acquisition temporalized ignorance, turning it into a temporary stage in a long development of which the present was merely a part.

3.4 Praise of Contingency: The Plurality of Worlds

John H. Brooke showed that between the belief that the world was made for man and the belief that it was not made for man, there was 'a long continuum of possibilities'—including that the world was 'chiefly, or only minimally, made for man' (Brooke 2000, 208–209). Huygens treatise *Cosmotheoros* provides a good example of these possibilities. Just at the beginning of the tract, Huygens emphasized that the statement that all things were created for man (*omnia propter hominum condita*) did not mean that all the stars were created only for human use or to be observed by human beings[52]:

---

[50] Hamou 2002, 44; Hamou 2008, 171–172, 176; Hamou 1999, vol. 1, 279–280; vol. 2, 86–90, 199–212.

[51] See e.g. Gassendi 1658, 2, vol. 333a, 456b; Glanvill 1885, 15, 56; Hooke 1665, fol. a1r-v.; Charleton 1654, 5; Boyle 1775, vol. 4, 50; Newton 1726, 529; Locke 1975, 555.

[52] Huygens 1888–1950, vol. 21, 687, English translation in Huygens 1698b, 7.

> Since then the greatest part of God's Creation, that innumerable multitude of Stars, is plac'd out of the reach of any man's Eye; and many of them, it's likely, of the best Glasses, so that they don't seem to belong to us; is it such an unreasonable Opinion, that there are some reasonable Creatures who see and admire those glorious Bodies at a nearer distance?

It was obvious to Huygens that there must be *spectatores* on other planets. Surely these were not beings like humans, but they were alive and using reason. 'For all this Furniture and Beauty the Planets are stock'd with seem to have been made in vain, without any design or end, unless there were some in them that might at the same time enjoy the Fruits, and adore the wise Creator of them.'[53] Huygens was thus able to apply the principle of witness in such a way that teleology of the relationship between the world and the stargazer was preserved without including anthropocentrism. The created cosmos needed an admiring gaze for its completion, but this gaze no longer needed to be human.

By the end of the seventeenth century, the idea that God wanted the universe filled with rational observers who could grasp the meaning, splendour, and rationality of God's work was not uncommon. Many seventeenth-century authors believed that without intelligent admirers, the beauty and order of the universe were meaningless. Due to their evidence-based admiration of God's work, humans and other stargazers became the crowning glory of creation.[54]

Fontenelle's *Entretiens sur la pluralité mondes* did not include natural theology, nor the principle of witness, nor *contemplatores caeli*. His presentation of the inhabited worlds was intended to make humans realize the contingency of their position in the universe and the constraints of the perspective with which they viewed the world. For Fontenelle, the particularity of the human perspective was a starting point for further investigation, not a pessimistic conclusion or a manifestation of epistemological helplessness.

From a cosmological point of view, the *Entretiens* were a comprehensible explanation of heliocentrism and Cartesian vortex theory. The cosmology that Fontenelle presented was barely visible, and therefore the readers had to free themselves from what they saw. Heliocentrism was only acceptable when the observer realized the relativity of his perspective. The readers had to take a rational standpoint in order to be able to admit that they were on an insignificant planet in one of the many solar systems. In her book, Isabelle Mullet explained in detail how Fontenelle repeatedly performed a multiplication of different perspectives to achieve this goal. Fontenelle portrayed how the same reality could appear quite different from various perspectives. As a result, a perspective that seemed natural to humans could be utterly confusing to other beings. According to Mullet, Fontenelle presented relativity as the universal framework of human knowledge (Mullet 2011, 23–39).

[53] Huygens 1888–1950, vol. 21, 687, 712–713; English translation in Huygens 1698b, 37.

[54] Borel 1657, 14–24, 37, 58; cf. Carré 1974, Guthke 1983, 91–105; Dick 1982, 69–101; Darmon 2002.

In creating images of the perspectivist conditioning of perception, Fontenelle drew on the full repertoire of ideas I have introduced in previous sections of this chapter. He moved from one abyss of infinity to another. He went over from the microworld to the human world and then to the universe to demonstrate how different beings were affected in their perception by their size, their location, and their physiology. Fontenelle described worms living in a single tree leaf who considered their tiny world 'immeasurable' and had no idea of their 'antipodes' on the other side of the leaf (Fontenelle 1991, 69–70). The pupae of silkworms, in turn, shared the movement of Earth without being aware of any movement (Fontenelle 1991, 35). By contrast, there was the perspective of the pre-Columbian cultures, who lived in primitive conditions and one day spotted huge ships dropping strange creatures ashore who held lightning in their hands (Fontenelle 1991, 53). Roses that live for only one day also viewed the world differently. Closed in their ephemeral one-day existence, roses could believe that the world was unchanging because they saw the same gardener for generations (Fontenelle 1991, 111).

Another perspective emerged with the extraterrestrial observers. Fontenelle tried to imagine the visual experience of extraterrestrials looking out from their locations into space: To philosophers from the Moon, our Earth would not appear motionless (Fontenelle 1991, 89–90). The inhabitants of Jupiter were unaware of the existence of the small and insignificant planet Earth until they built powerful telescopes (Fontenelle 1991, 68). In the Fontenellean universe, there were many subjectivities and many experiences, with no one being able to decide which was correct. Fontenelle even stated that there might be many senses in the universe, of which humans participated in only five. The extraterrestrials might have different senses with which they could perceive something different from humans (Fontenelle 1991, 72). The modern reader can think, for example, of ultraviolet radiation, which was discovered in the nineteenth century.

Naturally, the constraints of the human senses resulted in the limitations of human knowledge: 'Our sciences have certain limits which the human understanding has never been able to pass; there's a point at which they suddenly fail us. The rest is for other worlds, where some of what we understand is unknown.'[55] The knowledge of nature was thus also subject to a cosmic distribution, according to which inhabitants of various worlds always know only a certain part of nature according to the extent of their sensory equipment (including its technical improvement). Limited to their slice of reality, no community of rational beings could attain universal and absolute knowledge.

In his works, Fontenelle repeatedly challenged various thoughts of Pascal, as historians have noted, including the terrifying effects of the 'abysses of the infinite'.[56] At the beginning of the fifth dialogue of *Entretiens*, Fontenelle's philosopher informed the marquise that the stars were billions of miles away and that their

[55] Fontenelle 1991, 72; English translation in Fontenelle, 1990, 46; cf. Mullet's commentary (2011, 83–86).

[56] See more closely Mullet 2006, Mullet 2011, 50–56; cf. also Seidengart 2006, 538–542.

planetary systems were located around them. The marquise reacted to this presentation with mild dismay[57]:

> But here's a universe so large that I'm lost, I no longer know where I am, I'm nothing. What, is everything to be divided into vortices, thrown together in confusion? All this immense space which holds our Sun and our planets will be merely a small piece of the universe? [...] This confounds me – troubles me – terrifies me.

The marquise seemed to represent Pascal's libertine who felt anxiety when facing the contingent and vast universe. Fontenelle's philosopher did not. He reacted with a simple sentence 'this puts me at my ease' (Fontenelle 1991, 97). The philosopher explained that two reasons brought him to this answer, ontological and epistemological.

According to the philosopher, the small geocentric cosmos was limiting in ontological and epistemological sense. From the ontological point of view, the philosopher regarded the vast universe with its many vortices as an appropriate stage for the creative powers of nature, which only in such a vast space could produce the maximum degree of diversity as requested by the logic of the great chain of being. From the epistemological perspective, the geocentric universe was simply 'small and narrow'. In the immense universe, the philosopher breathed 'more freely' (*je respire avec plus de liberté*) (Fontenelle 1991, 98).

Fontenelle thus implied that the immense natural world corresponded to the free and inquisitive spirit that was not bound by deference to authority. The infinite universe was for the most part theoretically vacant, meaning that it was not explained by any prior theory or authority. Therefore it presented a natural challenge to human cognitive ambition, which could freely realize its potential in infinite objects.

Huygens's universe was even bigger, as seen in his estimates of the distances of stars. According to Huygens, nothing prevented God from creating infinitely many things beyond the visible stars that completely escape our understanding. However, similarly to Fontenelle, Huygens was not intimidated by the immensity of the universe: 'What is more admirable than the amazing vastness and magnificence of the world?' (Huygens 1888–1950, vol. 21, 813, 815–816).

The infinite universe did not arouse fear in Huygens, Fontenelle, or the translators and reviewers of their books.[58] In their view, a rational being did not feel distressed in the vast universe because it could at least approximately grasp it. The fear of an infinite universe, the vertigo of two abysses, the consciousness of human contingency—all these feelings were identified as prejudice arising from ignorance, from the absolutization of one's perspective, and from a misunderstanding nature that liked to realize its creative powers to the fullest extent. In a rational subject, whether terrestrial or extraterrestrial, the view of the immense universe aroused different reactions than fear and meaninglessness; specifically, it awakened wonder and curiosity.

---

[57] Fontenelle 1991, 98; English translation in Fontenelle 1990, 63.

[58] Huygens 1703, anonymous preface, fol. 4v; Huygens 1698a, anonymous preface, v-xxv; Anonymous 1698.

In the philosophical tradition since classical antiquity, the beginning of knowledge has been associated with the psychological states of wonder and curiosity, for which Daston and Park used the term cognitive passions (Daston and Park 2001, 304–305). Daston explained that during the seventeenth century there were significant changes in the relationship between wonder and curiosity. For philosophers of classical antiquity and the Middle Ages, wonder and curiosity were separate. Wonder, especially in the Aristotelian and Platonic traditions, was considered the legitimate beginning of knowledge, the origin and source of the natural human desire for knowledge. *Curiositas* was a vice: the proud transgression of the constraints of knowledge set for man by God. As Blumenberg demonstrated in his detailed history of curiosity, in the view of Christian theologians and apologetics, *curiositas* represented an interest in unnecessary things, such as nature. A curious human seeking knowledge of nature neglected the only thing that mattered: God and salvation.[59]

Using several examples, Daston and Park documented that around the mid-seventeenth century a merging of curiosity and wonder happened. They wrote that for Newton and the generation before him, wonder provoked curiosity. Amazement aroused the philosopher out of his idleness and brought his will and attention to a curious exploration of the world. According to Daston and Park, however, wonder and curiosity were again separated around the mid-eighteenth century. If wonder had traditionally been celebrated as the pre-eminent philosophical passion, for the Enlightenment it became a symptom of ignorance—only an uneducated and backward person wonders. In contrast, having been condemned for centuries, curiosity became an essential and legitimate quality of the diligent researcher (Daston and Park 2001, 326–327; Daston 2003, 79).

I do not think that Daston and Park were correct in all points, but they demonstrated the importance of legitimizing curiosity for establishing early modern philosophy, as Blumenberg did earlier in his *Die Legitimität der Neuzeit*. While wonder was always a kind of passive contemplation, curiosity gave knowledge dynamics by always seeking something new without ever being satisfied. The concept of a contingent world created space for the legitimate pursuit of curiosity, which no longer had to fear being accused of overstepping the boundaries of divine mysteries.

The defence of curiosity and delegitimizing of wonder can be found in Fontenelle's *Entretiens*. At the very beginning of the book, Fontenelle described the insatiable human desire for knowledge. All philosophy was based on an curious spirit (*l'esprit curieux*) and poor eyesight (*les yeux mauvais*). If we had perfect eyesight, we would immediately see whether the stars were the suns of other worlds. If we were less curious, we would not care for such knowledge. The 'trouble' was that we wanted to know more than we could see (Fontenelle 1991, 20). If humans could see things as they are, they would understand the world, but things do not appear to humans like that. Thus, for Fontenelle, motivation for cultivating philosophy resulted from 'trouble', from the disparity between human cognitive equipment

[59] On the history of curiosity, see Blumenberg 1996b, 263–528; Oberman 1974; Bös 1995.

and the human desire to understand nature. Natural philosophy arises as compensation for the weakness of our senses. Fontenelle thus summarized the early modern experience with limits of sensory experience in a few words: 'So true philosophers spend a lifetime not believing what they do see, and theorizing on what they don't see, and it's not, to my way of thinking, a very enviable situation.'[60]

Hamou compared Fontenelle's definition of philosophy with the beginning of Aristotle's *Metaphysics*, in which the Greek philosopher also said that 'all men by nature desire to know'.[61] In Aristotle, however, philosophy was associated with seeing in a positive sense. The pleasure of visual stimuli and the knowledge they bring attested to our natural desire for knowledge. Hamou explained that according to Aristotle, we philosophize because we see well while Fontenelle reversed this relationship. We philosophize because we see badly—and because our spirit was curious about things we cannot see (Hamou 1999, vol. 1, 13–14).

Fontenelle further developed this idea with a simile that occurred after the definition of philosophy. Fontenelle compared the study of nature to looking at the stage of a theatre. In contrast to the metaphor of *theatrum mundi,* Fontenelle replaced the vague image of a theatre with the specific building of the Paris Opera. He even referred to a specific theatre performance—Jean-Baptiste Lully's opera *Phaethon* (1683), which was performed in Versailles and later in other French cities (Mullet 2011, 152–156).

In Renaissance philosophy, the metaphor of the theatre suggested the cognitive accessibility of the world to man as the pinnacle of creation. Nature opened up to humans just as a stage opened to the glance of a visitor. Fontenelle fundamentally changed the meaning of the whole metaphor by explicating nature in terms of Cartesian mechanistic philosophy, which assumed that the mechanical properties of invisible corpuscles were the cause of visible phenomena. Fontenelle wrote, therefore, that by sight we saw only the scenery on the stage where a classical story took place. In ancient mythology, Phaethon was a figure who tried unsuccessfully to drive Helios's solar chariot. In Lully's opera, Phaethon's flight was arranged at the stage using hidden ropes and pulleys.

In Fontenelle's narrative, this mechanically mediated flight symbolized natural phenomena that have an unknown cause: 'the wheels and counter-weights that make everything move are hidden out of sight.'[62] Fontenelle then reproduced, with an excellent sense of literary shorthand, the views of several ancient philosophical schools on the causes of the fall, gravity, and the flight of bodies. He found their views ridiculous and singled out 'the great Descartes and some other moderns' who discovered the 'wires' carrying Phaethon (Fontenelle 1991, 21).

The effect of this literary simile was to disenchant nature through a mechanistic ontology. Fontenelle naturally did not want to say that natural philosophers knew everything, but that they were able to explain everything by means of an intelligible

[60] Fontenelle 1991, 20; English translation in Fontenelle 1990, 14.

[61] Aristotle, *Metaphysics*, I, 980a22-27; English translation in Aristotle 1995, vol. 2, 1552.

[62] Fontenelle 1991, 20; English translation in Fontenelle 1990, 12.

configuration of corpuscles. The philosopher could go backstage, behind the curtain, and there find that both opera and nature are well-functioning mechanisms. The philosopher eventually resembled the technician (*machinist*) behind the stage set who understood and drove the ropes and pulleys. Nature was not something to be admired, horrified of, or paralyzed by. Behind its spectacular processes laid the gears of a machine that was comprehensible to the human mind. When the marquise was introduced to the mechanical arrangement of nature, she replied that she held the world in much higher regard knowing that the world was like a watch: 'it's superb that, wonderful as it is, the whole order of nature is based upon such simple things.'[63]

In the words of the marquise, Fontenelle revived the Cartesian devaluation of admiration overlooked by Daston and Park. At the beginning of *Les Météores* (1637), Descartes had already expressed his dissatisfaction with admiration of nature because such a reaction led only to an imaginative and superstitious interpretation of nature. Causal interpretation was supposed to replace a naïve and pre-theoretical admiration of nature (Descartes 1996, vol. 6, 231). Similarly to Descartes, Fontenelle considered a simple observer (*spectateur*) the personification of an immature approach to nature that was based on wonder at the incomprehensible and often brought devising of superstitions or imaginary hypotheses. Maria Susana Seguin observed that Fontenelle confronted the *spectateur* with a *savant* who was not enchanted by nature but could rationally explain a miraculous or extraordinary phenomenon in terms of natural philosophy (Seguin 2016, 168–170).

Fontenelle thus reassessed the admiration that characterized the contemplative attitude of the stargazer. Anticipating the Enlightenment, Fontenelle regarded admiration and wonder as symptoms of stupidity and backwardness. Therefore, also in *Entretiens*, Fontenelle further developed the opera metaphor by claiming that humans had a distorted view of nature. 'Most cherish a false notion of mystery wrapped in obscurity. They only admire Nature because they believe she's a kind of magic, and the minute they begin to understand her they lose all respect for her.'[64]

In one of his later texts, Fontenelle explained that admiration for nature led to mere scepticism, which further caused contempt for natural science (*la science naturelle*). Fontenelle wanted to replace this view of nature with another one emphasizing the satisfaction brought about by knowledge: 'Nature is never so wonderful, nor so much esteemed, as when she is best understood' (Fontenelle 1989–2001, vol. 6, 47). Fontenelle thus continued to work with the category of wonder, but he changed the bearer of this characteristic. What aroused admiration was not nature itself, but the human ability to provide a natural explanation.

Such reflections meant the end of the idea of the stargazer. Fontenelle's marquise did not gaze at the world in sacred amazement but put a theoretical discourse between her senses and reality that stripped nature of its wonder and made it intelligible. Humans created their own world of natural explanations. Therefore,

---

[63] Fontenelle 1991, 22; English translation in Fontenelle 1990, 12.

[64] Fontenelle 1991, 22; English translation in Fontenelle 1990, 12; cf. Daston and Park 2001, 332–363.

Fontenelle was able to take the next step in developing the metaphor. The philosopher was sure that the marquise was ready for a non-magical interpretation of the world. The fantastic journey through the solar system began with symbolic words: 'I need only draw back the curtain and show you the world.'[65]

The curtain, as Mullet observed, was a symbolic interface between our perceptual world and the infinite universe (Mullet 2011, 28). The philosopher opened the curtain and revealed to the marquise the true nature of the world. Normally, humans could see only appearances. Philosophy was the only way to raise the curtain and emancipate humans from their naïve perspective. Fontenelle's Cartesian universe did not comprise Platonic harmonies, or Paracelsian sympathies but consisted of inert particles. At the same moment, it was not the mute and empty universe of Pascal's libertine. It was a universe brimming with intelligent life eager for knowledge. With admirable literary skill, Fontenelle wrote of 'the mutual curiosity with which the planets consider one another and ask among themselves, *What world is that? What people live on it?*'[66] Such a universe awakened joy at the omnipresence of reason, which eventually managed to overcome superstition by transforming it into scientific curiosity.

Huygens also broke with the ideal of the *contemplator caeli*. Humans and extraterrestrials were psychologically and epistemologically adapted to be stargazers but in a different form from that of Greek philosophy. In Huygens's view, God gave humans the diligence and acumen 'to make advances in the search of Nature' (Huygens 1888–1950, vol. 21, 687). Both diligence and reason were 'given' to humans by God, so they would be committing blasphemy if they did not use these gifts. The 'desire for knowledge and the love of knowledge' were not given to humans in vain. They were means by which humans could emancipate themselves from mere animal survival (Huygens 1888–1950, vol. 21, 737). Criticizing implicitly theological condemnations of curiosity, Huygens argued that if someone wanted to deny humans the right to extend the frontiers of knowledge, he would leave the human race in profound ignorance and in the futile horrors in which the barbarous nations were (Huygens 1888–1950, vol. 21, 823). In Huygens's homogenous universe, the same had to be true of extraterrestrials.

The fulfilment of the desire for knowledge happened through the development of natural philosophy.[67] For this reason, Huygens emphasized that the inhabitants of the planets did not limit themselves to observing and admiring natural phenomena but used their reason to explore the universe. Therefore, they also developed astronomy (Huygens 1888–1950, vol. 21, 731–733). According to Huygens, there was a great difference between those beings who idly (*otiose*) admired the beauty of the Sun and those who measured the distances of bodies and understood the laws of their movements. Similarly, there was a difference between those who merely

[65] Fontenelle 1991, 22; English translation in Fontenelle 1990, 12.

[66] Fontenelle 1991, 90; English translation in Fontenelle 1990, 57.

[67] Huygens 1888–1950, vol. 21, 731; English translation in Huygens 1698b, 60.

admired the variety and activities of animals and those who grasped the arrangement of their organs (Huygens 1888–1950, vol. 21, 731).

Similarly to Fontenelle, Huygens reconceptualized the notion of wonder beyond its simple rejection typical of the Cartesian tradition. In his view, knowledge of nature had been freed from its subordinate function of moral improvement and salvation and become a legitimate end in itself. A rational being did not achieve happiness by passive contemplation, but through an active understanding of the universe. Huygens argued that humans were brought to astronomy and natural philosophy by wondering at eclipses and the movements of the stars. Sciences must then have developed very early on the planets with many satellites, such as Jupiter, where eclipses of the moons occurred almost constantly (Huygens 1888–1950, vol. 21, 733). Rather than being an expression of philosophical sensitivity to the splendour of nature, wonder became a natural part of empirical science, as it turned into an opportunity for the active deployment of observational methods.

Huygens's work exemplified Daston's reflection on 'the intricate minuet of wonder and curiosity' (Daston and Park 2001, 304) in the seventeenth and eighteenth centuries. For Huygens, wonder was not yet the mark of the ignorance of the simpleton as it would become for the Enlightenment. However, curiosity was no longer the manifestation of sinful pride as it was for Christian theologians and devote philosophers. In Huygens's view, the two cognitive passions still supported each other.

In sum, both Fontenelle and Huygens demonstrated that the idea of an immeasurable universe containing two infinite abysses need not have brought about either sceptical resignation, fideism, or condemnation of curiosity. Natural philosophers at the end of the seventeenth century could see the universe as a reservoir of potential objects of knowledge on which the desire for knowledge common to all rational beings could be exhausted. The realization of this desire took a double form. It brought about the overcoming of prejudices and errors that were the legacy of the past. Furthermore, it resulted in the anticipation of further discoveries in the future. The curious desire for knowledge expressed the new dynamic temporal regime of knowledge since it related to both the past and the future.

The preceding chapters demonstrated that ontological, cosmological, and anthropological contingency was an essential condition for the formation of the prescriptive idea of scientific progress. Natural philosophers reasoned that reality and the future contained an infinite reservoir of possible objects of knowledge that guaranteed the possibility of further expansion of knowledge. Contingency provided the possibility of perfectibility because that which is contingent need not always remain the same and can be improved. Such improvement could concern perception, technology, or the extent of empirical knowledge. Contingency also abolished the natural intelligibility of nature assumed by Renaissance philosophy and brought scepticism regarding knowledge of empirical nature. This scepticism, however, could be temporalized by making ignorance temporary. Further, contingency implied the absence of permanent epistemological boundaries that were naturally or supernaturally determined. This lack of metaphysical warrants of human knowledge was interpreted as epistemological freedom resulting in knowledge expansion.

Contingency also removed the belief in the possibility of definitive and causal knowledge, *scientia*. This created the need for a new type of knowledge that was appropriate to human cognitive abilities and the contingency of their relation to reality. Finally, the recognition of contingency gave the new philosophy and sciences a self-confidence that resulted both from the pride of overcoming prejudice and from the fact that all knowledge was earned and acquired autonomously without teleological or transcendental underpinnings. All these changes resulted in the introduction of a new methodological trait of natural philosophy, which I have termed the temporal regime of knowledge.

Bibliography

Anonymous. 1698. Article IX: Christiani Hugenii Kosmotheoros. In *Histoire des ouvrages des Savans*, Mois de Sept. Octob. Nov. Dec. 1697. Janv. Fevr. Mars 1698, 228–242. Rotterdam.

Aristotle. 1995. *The complete works of Aristotle*, ed. Jonathan Barnes. 2 vols. Princeton: Princeton University Press.

Arnauld, Antoine, and Nicole, Pierre. 1854. *La Logique de Port-Royal*, ed. Charles Jourdain. Paris: Hachette.

Basse Eriksen, Christopher. 2018. *Beneath the visible: Nature, and the sub-visible world in early modern microscopy*. Aarhus University. Ph.D. thesis.

Ben-Yami, Hanoch. 2015. *Descartes' philosophical revolution: A reassessment*. New York: Palgrave.

Berkeley, George. 2009. *Essay on a new theory of vision*. In George Berkeley, *Philosophical writings*, ed. Desmond M. Clarke, 1–65. Cambridge: Cambridge University Press.

Blair, Ann. 1997. *The theatre of nature. Jean Bodin and renaissance science*. Princeton: Princeton University Press.

Blumenberg, Hans. 1966. Contemplator coeli. In *Orbis scriptus. Dmitri Tschizewsky zum 70. Geburstag*, ed. Dietrich Gerhardt, 113–124. Munich: Fink.

———. 1986. *Lebenszeit und Weltzeit*. Frankfurt am Main: Suhrkamp.

———. 1987. *The genesis of the Copernican world*, translated by Robert M. Wallace. Cambridge: MIT Press.

———. 1996a. *Die Genesis der kopernikanischen Welt*. Frankfurt am Main: Suhrkamp.

———. 1996b. *Die Legitimität der Neuzeit. Erneuerte Ausgabe*. Frankfurt am Main: Suhrkamp.

———. 2002. Das Fernrohr und die Ohnmacht der Wahrheit. In Galieo Galilei, *Sidereus Nuncius. Nachricht von neuen Sternen*, ed. Hans Blumenberg, 7–75. Frankfurt am Main: Suhrkamp.

Bodin, Jean. 1597. *Universae naturae theatrum*. Francofurti: Wechelius.

Borel, Pierre. 1655. *De vero telescopii inventore*. Hagae: Vlacq.

———. 1656. *Observationum microscopicarum centuria*. Hagae: Vlacq.

———. 1657. *Discours nouveau prouvant la pluralité des Mondes*. Genève: publisher anonymous.

Bös, Günther. 1995. *Curiositas. Die Rezeption eines antiken Begriffes durch christliche Authoren bis Thomas von Aquin*. Paderborn: Schöningh.

Bovillus, Carolus. 1510. *Liber de sapientiae*. In Carolus Bovillus, *Liber de intellectu. Liber de sensibus. Liber de generatione. Libellus de nihilo. Ars oppositorum. Liber de sapiente. Liber de duodecim numeris. Philosophicae epistulae. Liber de perfectis numeris. Libellus de mathematicis rosis. Liber de mathematicis corporibus. Libellus de mathematicis supplementis*. Paris: Estienne.

Boyle, Robert. 1775. *The works*, ed. Thomas Birch. 6 vols. London: Rivington.

Brague, Rémi. 1994. Geozentrismus als Demütigung des Menschen. *Internationale Zeitschrift für Philosophie* 1: 2–25.

———. 2005. *Die Weisheit der Welt. Kosmos und Erfahrung im westlichen Denken*. Munich: C. H. Beck.

Brahe, Tycho. 1913–1929. *Tychonis Brahe Dani opera omnia*, ed. John L. Dreyer, vol. 10 vols. Copenhagen: Gyldendal.

———. 1929. On mathematical studies, translated by R. H. Coon, *Popular Astronomy* 37: 311–320.

Brooke, James H. 2000. Wise men nowadays think otherwise: John Ray, natural theology and the meaning of anthropocentrism. *Notes and Records of the Royal Society of London* 54: 199–213.

Bruyère, Jean, de la. 1885. *The characters*, translated by Henri van Laun. London: Nimmo.

Cardano, Girolamo. 1580. *De rerum varietate*. Lugduni: Michel.

Carré, Marie-Rose. 1974. A man between two worlds: Pierre Borel and his *Discours nouveau prouvant la pluralité des mondes* of 1657. *Isis* 65: 322–335.

Caston, Victor. 2005. The spirit and the letter: Aristotle on perception. In *Metaphysics, soul, and ethics in ancient thought. Themes from the work of Richard Sorabji*, ed. Ricardo Salles, 245–320. Oxford: Clarendon Press.

Charleton, Walter. 1654. *Physiologia Epicuro-Gassendo-Charltoniana*. London: Newcomb.

Cicero. 1967. *De natura deorum. Academica*, translated by H. Rackham. Cambridge, Mass./London: Harvard University—Heinemann.

Clavius, Christopher. 1606. *In Sphaeram Ioannis de Sacro Bosco Commentarius*. Romae: Gellius.

Cohen, Bernard I. 1940. Røemer and the first determination of the velocity of light (1676). *Isis* 31: 327–379.

Comenius, Johannes A. 1964. *De rerum humanarum emendatione consultatio catholica*. 2 vols. Academia: Pragae.

Cordemoy, Géraud de. 1666. *Le discernement du corps et de l'ame en six discourse*. Paris: Lambert.

Croll, Oswald. 1620. *Basilica Chymica*. Francofurti: Tampachius.

Darmon, Jean-Charles. 2002. L'imagination de l'espace entre argumentation philosophique et fictions. *Études Littéraires* 34: 217–239.

Daston, Lorraine. 2003. Die kognitiven Leidenschaften: Staunen und Neugier im Europa der frühen Neuzeit. In Lorraine Daston, *Wunder, Beweise und Tatsachen. Zur Geschichte Rationalität*, 77–97. Frankfurt am Main: Fischer.

Daston, Lorraine, and Katherine Park. 2001. *Wonders and the order of nature*, 1150–1750. New York: Zone Books.

Descartes, René. 1984–1991. *The philosophical writings*, 3 vols. translated by John Cottingham, Robert Stoothoff, Dugald Murdoch, and Athony Kenny. Cambridge: Cambridge University Press.

———. 1996. *Œuvres de Descartes*, ed. Charles Adam and Paul Tannery, vol. 11 vols. Paris: Vrin.

Dick, Steven J. 1982. *Plurality of worlds. Origins of the extraterrestrial life debate from Democritus to Kant*. Cambridge: Cambridge University Press.

Fontenelle, Bernard, de. 1727. *Lives of the French, Italian and German philosophers, late members of the royal academy of sciences in Paris*. London: Innys.

——— 1989–2001. Bernard de Fontenelle, *Œuvres complètes*, ed. Alain Niderst, 9 vols. Paris: Fayard.

———. 1990. *Conversations on the plurality of worlds*, translated by H. A. Hargreaves. Berkeley: University of California Press.

———. 1990–2001. *Œuvres complètes*, ed. Alain Niderst, 9 vols. Paris: Fayard.

———. 1991. Entretiens sur la pluralité des mondes. In Bernard de Fontenelle, *Œuvres complètes*, ed. Alain Niderst, vol. 2. Paris: Fayard.

———. 2016. *Digression sur les Anciens et les Modernes et autres textes philosophiques*, ed. Sophie Audidière. Paris: Garnier.

Fontenelle, Bernard, de, and Louis Godin. 1733. *Histoire de l'Académie Royale des Sciences, depuis son Etablissement en 1666 jusqu'à 1699*. 2 vols. Paris: Martin, Coignard & Guerin.

Galilei, Galileo. 1890–1909. *Edizione nazionale delle opere di Galileo Galilei*, ed. Antonio Favaro, 20 vols. Firenze: Barbèra Editore.

———. 1960. *The assayer*. In Galileo Galilei, Horatio Grassi, Mario Guiducci, and Johannes Kepler, *Controversy on the comets of 1618*, translated by Stillman Drake and C. D. O'Malley. Philadelphia: University of Pennsylvania Press.

———. 1989. *Sidereus nuncius, or the sideral messenger*, translated by Albert van Helden. Chicago/London: The University of Chicago Press.

Gassendi, Pierre. 1658. *Opera omnia*. 6 vols. Lugduni: Anisson.

Gaukroger, Stephen. 2002. *Descartes' system of natural philosophy*. Cambridge: Cambridge University Press.

Gessner, Conrad. 1551–1555. *Historia animalium*. 3 vols., Tiguri: Froschoverus.

Glanvill, Joseph. 1885. *Scepsis scientifica, or vanity of dogmatizing*, ed. James Owen. London: Kegan Paul, Trench & Company.

Gregorić, Pavel. 2019. Aristotle's perceptual optimism. *Croatian Journal of Philosophy* 19: 543–560.

Groh, Dieter. 2003. *Schöpfung im Widerspruch: Deutungen von der Natur des Menschen von der Genesis bis zur Reformation*. Frankfurt am Main: Suhrkamp.

Grynaeus, Stephan. 1533. Praefatio. In Euclid, *Euclidis Elementa*. Basel, fols. a1r-5r.

Guthke, Karl S. 1983. *Der Mythos der Neuzeit. Das Thema der Mehrheit der Welten in der Literatur – und Geistesgeschichte von der kopernikanischen Wende bis zur Science Fiction*. Bern/München: Francke.

Hallyn, Fernand. 1992. Introduction. In Galileo Galilei, *Le Messager des étoiles*, 11–130. Paris: Seuil.

Hamou, Philippe. 1999. *La mutation du visible. Essai Sur la portée épistemologique des instruments d'optique au XVIIe siècle*. 2 vols. Paris: Septentrion.

———. 2002. *Voir et connaître à l'âge classique*. Paris: Presses Universitaires de France.

———. 2004. Microscopes et minima visuels. Berkeley critique de l'autopsie instrumentale. In *Science et épistémologie selon Berkeley*, ed. Sébastien Charles, 103–125. Québec: Canadian Philosophical Association.

———. 2008. La nature est inexorable. Pour une reconsidération de la contribution de Galilée au problème de la connaissance. *Galilaeana* 5: 149–177.

Hooke, Robert. 1665. *Micrographia, or some physiological descriptions of minute bodies made by magnifying glasses*. London: Martyn and Allestry.

———. 1705. *Posthumous works*. London: Waller.

Hortensius, Martin. 2006. The oration on the dignity and the usefulness of the mathematical sciences of Martinus Hortensius (Amsterdam, 1634): Text, translation and commentary, edited and translated by A. Immhausen, and Volker R. Remmert. *History of Universities* 21: 71–150.

Huygens, Christiaan. 1698a. Nouveau Traité de la Pluralité des Mondes, traduit par. M. D. ***, Amsterdam: Moreau.

———. 1698b. *The celestial worlds discover'd or, conjectures concerning the inhabitants, plants and production, of the worlds in the planets*. London: Childe.

———. 1703. *Cosmotheoros oder Welt-Betrachtende Muthmassungen*. Leipzig: Lanskisch.

———. 1888–1950. *Œuvres complètes*. 22 vols. La Haye: Nijhoff.

Jansenius, Cornelius. 1642. *Discours de la reformation de l'homme interieur*. Paris: Camusat.

Jasiński, René. 1933. Sur les deux infinis de Pascal. *Revue d'histoire de la philosophie et d'histoire générale de la civilisation* 1: 134–159.

Jessenius, Johannes. 1601. *Anatomiae Pragae historia*. Witterbergae: Scheurlich.

Kepler, Johannes. 1938–2009. *Gesammelte Werke*, ed. M. Caspar, V. Bialas, et al., 25 vols. München: C. H. Beck.

Kusukawa, Sachiko. 1995. *The transformation of natural philosophy. The case of Philip Melanchton*. Cambridge: Cambridge University Press.

Leeuwenhoek, Anthoni van. 1939–1999. *Alle de brieven*. 15 vols. Amsterdam: Swets and Zeitlinger.

Leinkauf, Thomas. 2009. *Mundus combinatus: Studien zur Struktur der barocken Universalwissenschaft am Beispiel Athanasius Kirchers SJ (1602–1680)*. Berlin: Akademie Verlag.

Lindberg, David Charles, and Nicholas H. Steineck. 1972. The sense of vision and the origins of modern science. In *Science, medicine and society in the renaissance. Essays to honor Walter Pagel*, ed. Allen George Debus, vol. 1, 29–45. New York: Heinemann.

Locke, John. 1975. *Essay concerning human understanding*. ed. Peter H. Nidditch. Oxford: Clarendon Press.

Lüthy, Christoph. 1996. Atomism, Lynceus, and the fate of seventeenth-century microscopy. *Early Science and Medicine* 1: 1–27.

Malebranche, Nicolas. 1979. *De la recherche de la vérité*. In Malebranche, Nicolas. *Oeuvres*, ed. Geneviève Rodis-Lewis, vol. 1. Paris: Gallimard.

———. 1997. *The search after truth*, translated by Thomas M. Lennon – Paul J. Olscamp. Cambridge: Cambridge University Press.

Mästlin, Michael. 1578. *Observatio & Demonstratio Cometae Aetherei*. Tubingae: Gruppenbach.

———. 1588. *Epitome astronomiae*. Tubingae: Gruppenbach.

Mullet, Isabelle. 2006. Vertige et tourbillons: les *Entretiens sur la pluralité des mondes* et la révolution du point du vue. *Revue Fontenelle* 4: 26–40.

———. 2011. *Fontenelle ou la machine perspectiviste*. Paris: Honoré Champion.

Newton, Isaac. 1726. *Philosophiae naturalis principia mathematica*. Londini: Innys.

Oberman, Heiko. 1974. *Contra vanam curiositatem. Ein Kapitel der Theologie zwischen Seelenwinkel und Weltall*. Zürich: Theologischer Verlag.

Paracelsus. 1924–1933. *Sämtliche Werke*, ed. Karl Sudhoff, 14 vols. Munich/Berlin: Oldenbourg.

Pascal, Blaise. 1995. *Pensées and other writings*, translated by Honor Levi. Oxford: Oxford University Press.

———. 1998. *Œuvres complètes*, ed. Michel Le Guern, 2 vols. Paris: Gallimard.

Peucer, Caspar. 1563. *Elementa doctrinae de circulis coelestibus*. Vitebergae: Lufftius.

Pfeiffer, Jens. 2001. *Contemplatio Caeli. Untersuchungen zum Motiv der Himmelsbetrachtung in lateinischen Texten der Antike und des Mittelalters*. Hildesheim: Weidmann.

Pico della Mirandola, Giovanni. 2019. *Oration*. In *Magic and the dignity of man. Pico della Mirandola and his oration in modern memory*, ed. Brian Copenhaver, 459–482. Cambridge, Mass./London: The Belknap Press of Harvard University Press.

Power, Henry. 1664. *Experimental philosophy*. London: Roycrofe.

Probst, Peter. 2021. Spectator coeli. In *Historisches Wörterbuch der Philosophie*, ed. Joachim Ritter and Karl Gründer, vol. 9, 1350–1355. Basel/Stuttgart: Schwabe.

Ptolemy, Claudius. 1998. *Ptolemy's Almagest*, translated and ed. Gerald J. Toomer. Princeton: Princeton University Press.

Regiomontanus, Johannes. 1544. *Scripta Clarissimi Mathematici M. Ioannis Regiomontani*. Nurembergae: Montanus and Neuber.

Régis, Pierre-Sylvain. 1690. *Système de philosophie, contenant la logique, la métaphysique, la physique et la morale*. Paris: Anisson.

Reinhold, Erasmus. 1551. *Prutenicae tabulae coelestium motuum*. Tubingae: Ulricum.

Rheita, Schyrlaeus. 1645. *Oculus Enoch et Eliae sive radius sidereomysticus*. 2 vols. Antverpiae: Verdussius.

Rheticus, Joachim. 1542. *Orationes duae prima de astronomia & geographia altera de physica*. Nurembergae: Petreius.

Riccioli, Giambattista. 1651. *Almagestum novum*. Bononiae: Benatius.

Roberts, Frank. 1694. Concerning the distance of the fixed stars. *Philosophical Transactions* 18: 101–103.

Roelants, Nienke. 2012. The physical status of astronomical models before the 1570s: The curious case of Lutheran astronomer Georg Joachim Rheticus. *Science and Theology* 10: 367–390.

Rohault, Jacques. 1671. *Traité de physique*. Paris: Thierry.

Rohault, Jacques, and Samuel Clarke. 1735. *System of natural philosophy illustrated with Dr. Samuel Clarke's notes taken mostly out of Sir Isaac Newton's philosophy*. 2 vols. London: John/Knapton.

Rondelet, Guilaume. 1558. *L'Histoire entiere des poissons*. Lion: Bonhome.

Scheiner, Christoph. 1630. *Rosa ursina sive Sol*. Bracciani: Phaeus.

Seguin, Maria Susana. 2016. Présentation: Sur la lumière septentrionale. In Bernard de Fontenelle, *Digression sur les Anciens et les Modernes et autres textes philosophiques*, ed. Sophie Audidière, 168–170. Paris: Garnier.

Seidengart, Jean. 2006. *Dieu, l'univers et la sphére infinie. Penser l'infinité cosmique à l'aubé de la science classique*. Paris: Albin Michel.

Seneca, Lucius, Annaeus. 2010. *On leisure*, translated Gareth D. Williams. In Lucius Annaeus Seneca, *Hardship & Happiness*. Chicago/London: The University of Chicago Press.

Smith, Mark A. 2018. *From sight to light. The passage from ancient to modern optics*. Chicago/London: The University of Chicago Press.

Specht, Robert. 2009. Experiences and hypotheses: Opinions within Locke's realm. In *The significance of the hypothetical in the natural sciences*, ed. Michael Heidelberger and Gregor Schiemann, 38–58. Berlin/New York: De Gruyter.

Stadius, Johannes. 1560. *Tabulae Bergenses aequabilis et apparentis motus orbium coelestium*. Coloniae: Brickmann.

Stoffel, Jean-François. 1998. La révolution copernicienne et la place de l'homme dans l'Universe. Étude programmatique. *Revue philosophique de Louvain* 96: 7–50.

———. 2001. Géocentrisme, héliocentrisme, anthropocentrisme: quelles interactions? *Scientiarum Historia* 27: 77–92.

———. 2002. La révolution copernicienne responsable du 'désenchantement du monde'? L'exemple des analogies solaires. *Revue belge de philologie et d'histoire* 80: 1189–1224.

Taub, Lisa. 1993. *Ptolemy's universe. The natural philosophical and ethical foundations of Ptolemy's astronomy*. Chicago: Open Court.

Tyard, Pontus, de. 1557. *L'univers ou discours des parties, et de la nature du monde*. Lion: Tournes.

Vanini, Giulio. 1615. *Amphiteatrum aeternae providentiae*. Lugduni: Harsy.

Wilson, Catherine. 1988. Visual surface and visual symbol: The microscope and the occult in early modern science. *Journal of the History of Ideas* 49: 85–108.

———. 1995. *The invisible world. Early modern philosophy and the invention of the microscope*. Princeton: Princeton University Press.

Witthaus, Jan-Henrik. 2005. *Fernrohr und Rhetorik. Strategien der Evidenz von Fontenelle bis La Bruyère*. Heidelberg: Winter.

Wolf-Devine, Celia. 2000. Descartes' theory of visual spatial perception. In *Descartes' natural philosophy*, ed. Stephen Gaukroger, John Schuster, and John Sutton, 504–522. London/New York.

Wren, Charles. 1721. *Parentalia, or memoirs of the family of the Wrens*. London: Osborn.

Chapter 4
The Temporal Regime of Knowledge

Contents

Abstract The Académie des sciences in Paris started to work in December 1666. Its members put their research to the temporal regime of knowledge by which ignorance was understood as temporary. The inexhaustibility and diversity of nature necessitated the extension of knowledge over a longer time horizon: research therefore had to be long-term and fundamentally unconstrained. Because of the dependence on developing observational and experimental instruments, the limits of human knowledge were not fixed and absolute, but temporary and relative, implying a constant further expansion of knowledge in the future. The members of the Académie des sciences mostly recognized that mechanist explanations of natural processes were essentially hypothetical because of the unobservability of the constitutive parts of nature (i.e., corpuscles). Hypotheses were expected to be temporary and provisional. The traditional Aristotelian ideal of *scientia* was thus replaced with hypothetical knowledge that did not assert its definitiveness. The Secretary of the Académie, Bernard de Fontenelle, generally referred to the epistemological practice of academicians as a sage Pyrrhonism that included several methodological features: the pursuit of a continuous accumulation of data without interpretation; opposition to hasty conclusions and supposedly definitive systems; the suspension of judgment; the slowness and gradualness of research; and the advancement of the sciences from specific facts towards general principles. This methodology left open the question of whether the refinement of scientific knowledge would lead to definitive truth or continue without the possibility of reaching perfection.

D. Špelda, *The Origins of the Idea of Scientific Progress*, International Archives of the History of Ideas Archives internationales d'histoire des idées 250,
https://doi.org/10.1007/978-3-031-60526-0_4

Keywords Académie des sciences · Bernard de Fontenelle · Hypotheses · Mechanism · Scientific scepticism · Idea of scientific progress

The contingency of nature caused the emergence of new methodological characteristics that entered natural philosophy in the second half of the seventeenth century: the impossibility of concluding the knowledge process because of the inexhaustibility of nature; the limitation of natural philosophy to hypotheses; the gathering and transmission of facts without a theoretical framework; the abstention from hasty judgments; the avoidance of systems and definite answers; the graduality and slowness of knowledge advancement; the sceptical suspension of judgment. I understand these characteristics to be essential components of the temporal regime of knowledge. Turning knowledge into a temporal regime allowed natural philosophy not to get lost in the infinite abysses of the universe by making human ignorance temporary and relative. I understand the introduction of the temporal regime of knowledge as further condition of possibility of the idea of scientific progress.

In this chapter, I present the description of the temporal regime of knowledge by breaking it down into the individual characteristics mentioned above. Although I also refer to other representatives of early modern philosophy, my primary sources are the works produced at the Académie des Sciences in Paris. As the Académie is rather neglected theme in early modern research I want to begin with a short sketch of the history of institution based on works of Roger Hahn, Alice Stroup, David J. Sturdy, Marie-Jeanne Tits-Dieuaide, Guy Picolet, and Simone Mazauric.[1]

4.1 Académie des Sciences

From about the 1660s onwards, various scholars appealed to the king, and to the state minister Jean-Baptiste Colbert, to support natural philosophy and sciences in the form of an institution that would carry out an empirical programme similar to Bacon's idea of long-term research (Sorbrière 1917). From an organizational and political point of view, the diplomat and scholar Jean Chapelain, the mathematician and royal librarian Pierre de Carcavi, and the brothers Claude and Charles Perrault played key roles. It was with them that Colbert consulted on the shape of the new institution. Huygens was initially meant to be the Académie's leading figure, but he was outshone in the 1670s by Cassini who was able to publish more quickly about more spectacular discoveries (Van Helden 1996).

By founding the Académie, Colbert hoped to achieve several goals. The political goal was to strengthen the prestige of the king and to extend the royal authority in the intellectual sphere of French society. The economic goal was to develop useful

[1]Mazauric 2007, 2009, 143–154; Dew 2015; Donato 2015; Picolet 2021; Tits-Dieuaide 1998; Sturdy 1995, 17–95.

technologies to serve the nation, such as maps and aqueducts. However, administrative pragmatism and utilitarianism were not the only reasons for founding the Académie. Mazauric (2007, 43) highlighted letter from Colbert's correspondence concerning the founding of the Académie in which an anonymous writer stressed that the 'destruction of ignorance' was an appropriate task for the monarch and his minister. According to this letter to Colbert from 1665, the fight against ignorance was to include the following items: 'to banish from the sciences all prejudices, to rely only on experience, to find something certain in sciences, to remove from them all chimeras and to open to those who will cultivate the sciences in the future an easy path to truth, as much as God allowed humans to know them for their utility'.[2]

Mazauric rightly pointed to the normative ambition and demarcation problem of this description. From the beginning, alchemy and astrology were excluded from the Académie's programme, and thus from the official discourse of natural philosophy, which was sanctified by the authority of the royal administration (Mazauric 2007, 43). For our investigation, it is important that the temporal orientation of the whole project was evident in the quoted sentences since there was an explicit emphasis on gradual and long-term research. From the outset, the Académie was planned as a bearer of knowledge that would not be subject to the demands of immediate utility and would also formulate epistemological standards in the long run and prepare for future discoveries.

The first meeting of the Académie took place in December 1666, in the King's library in Rue Vivienne. Members of the Académie's mathematical section, including Huygens and Adrien Auzout, drew up specific research plans to be carried out in astronomy and other mathematical fields.[3] These plans were partly preserved in the letters, manuscripts, and registers of the Académie. An important part of them was the request for the establishment of a royal astronomical observatory, which the king and Colbert granted. The foundation stone of the Paris Observatoire was laid by mathematicians from the Académie in 1667 on the day of the summer solstice.[4] The majestic building of the Observatoire, which was completed in 1672, was originally planned as the headquarters of the Académie and therefore included rooms for performing experiments and dissections. However, the Académie was never moved to the Observatoire building. The Observatoire served the astronomers, while the rest of the academicians worked and lived in the Royal Library buildings, and from 1699 onwards in the Louvre.

The early Académie was small. At its foundation, it had 15 academicians and five students (*élèves*). Carcavi was the director; Jean-Baptiste Du Hamel acted as the secretary. The academicians were divided into two sections. The first consisted of

---

[2]Colbert 1888, 513; cited by Mazauric 2007, 43.

[3]Auzout 1665, 'Au Lecteur' (unpaginated); Auzout 1664, 'Au roy' (unpaginated); cf. McKeon 1965. Picard's version of the programme was published by Wolf (1902, 203–204). Huygens made only a list of propositions including also natural philosophy; see Huygens 1888–1950, vol. 19, 255–257, cf. vol. 8, 196–198.

[4]Wolf 1902, 11; Jelaco 2016.

mathematicians such as Huygens, Cassini, and Gilles de Roberval. The second group was made up of *physiciens* such as Claude Perrault, Claude Bourdelin, and Samuel Duclos. They were concerned with 'physics' in a very broad sense, covering both traditional disciplinary fields: natural philosophy and natural history. Members of this group were engaged in empirical disciplines that would nowadays be classified as anatomy, physiology, chemistry, botany, or zoology. Representatives of the 'physical' section emphasized empirical research and data collection. The members of the mathematical section worked on mathematics (for example, Marquis L'Hopital developed infinitesimal calculus) or concentrated on developing the mathematical apparatus of mechanics, thus laying the foundations of the field later called rational mechanics. The most important representative of mechanics in the early Académie was Fontenelle's friend and roommate Pierre Varignon, who became a member of the Académie in 1688.

Although the two sections were considered thematically and methodologically distinct, we should not assume that there was a separate section of mathematicians that was devoted only to abstract research while the natural philosophers and natural historians carried out empirical research. In fact, the two groups were considered two offshoots of the indivisible body of the Académie. The connection between the two groups existed on several levels. On a practical level, cooperation between the two sections was manifested in joint meetings held twice a week. On Wednesdays, they discussed problems in mathematics; on Saturdays, they dealt with 'physics'. Thus, although mathematics and mechanics can be said to be oriented towards rational and abstract procedures, in reality mathematicians had a perfect overview of the experimental work of their colleagues and participated in the methodological values and decisions of the Académie. The idea that the two sections were separated could be also called in question by the fact that astronomy and optics, which were strongly empirically and experimentally oriented, were counted among the mathematical disciplines. Finally, some academicians, despite their formal classification in one of the two sections, moved freely between mathematical and empirical types of research. For example, Philippe de la Hire was involved in mathematics, astronomy, but also meteorology, aqueduct building or physiology.

Thus, when I describe the temporal regime of knowledge in the text of this chapter, I work mostly with evidence and examples from empirical disciplines because the collection and evaluation of empirical data has always taken place over time. But this does not mean that members of the mathematics section think differently about the role of time in research methodology. Scholars like Huygens, Varignon, and L'Hopital had a lot of experience with experimental and empirical research because they had done it themselves or saw it at the meetings of the Académie every week.

Colbert generously supported the Académie. Academicians received regular pensions (at least during Colbert's life) for their membership in the Académie, although the amount varied. Cassini and Huygens received the most money; other academicians had significantly lower pensions. When Colbert died in 1683, François-Michel le Tellier, Marquis de Louvois, who was minister of war, took over responsibility for the Académie. Louvois was not nearly as interested in the

Académie's activities as Colbert. He did not care much for theoretical scientific research and expected the academicians to produce practical results, such as the irrigation of Versailles and a cure for the ailing king.

In 1691, Louvois died. After his death, the minister Louis Phélypeaux de Pontchartrain became the protector of the Académie, appointing his nephew Abbé Jean-Paul Bignon as the director of the institution. In the 1680s and 1690s, the budget of the Académie was drastically reduced because of the wars, the economic situation, and the policies of the new ministers, which differed from those of Colbert. Stroup has charted the Académie's funding in great detail.[5] I give just one example from her analysis, which concerns the Observatoire and well illustrates the differences in financial support: 'Pontchartrain spent on average only 319 livres a year on the Observatoire, in contrast with 39,650 livres under Colbert and 1,371 under Louvois' (Stroup 1990, 57). In the 1680s and 1690s, the pensions of the academicians were also reduced or paid with delay.

The difficult period of the Académie came to an end in 1699, when Pontchartrain and Bignon carried out a comprehensive reform of the institution. At the beginning of 1699, the president of the society, Bignon, presented a new system of rules governing research at the Académie. The reform, to which I return in the fifth chapter, mandated that the Académie issue regular reports on its activities. Fontenelle, the new secretary of the institution, was entrusted with their compilation.

In the following decades, Fontenelle prepared volumes acquainting the public with the work of the Académie each year. The volumes were entitled *Histoire de l'Académie royale des sciences avec le Mémoires de Mathématique & de Physique, pour la même Année: Tirez des Registres de cette Académie* (further as *HARS*). The first volume, covering the academic year 1699, was not published until 1702. Subsequent volumes were also published two to three years after the academic year they described. The last volume prepared by Fontenelle was the volume describing the academic year 1740, which appeared in 1742. Subsequent volumes were published by Fontenelle's successors in the capacity of secretary.

In the first volume for 1699, Fontenelle published the new administrative rules and introduced the volume itself with a preface explaining the nature of scientific research at the Académie. Fontenelle's preface was in many ways an extraordinary text because it successfully summarized the methodological tenets of research at the Académie at the turn of the century. For this reason, the text has been published as an important part of Fontenelle's philosophical work under the title *Préface sur l'utilité des mathématiques et de la physique*.

Each volume of *HARS* consisted of two parts with separate pagination. The first part was referred to as *histoire*; it was written by Fontenelle, although the author was not mentioned. In this part, Fontenelle acquainted the readers with the research of academicians, sometimes supplementing sober descriptions with his own comments

[5] Stroup 1990, 46–59. Stroup described the budgets of the Académie and the Observatoire in great detail in the tables at the end of the book, see 243–252; cf. Wolf 1902, 209–215; Shank 2018, 46–77.

(further quoted as Fontenelle 1702–1742). The second (and larger) part of each volume consisted of texts by the academicians themselves. This part was entitled *Mémoires de mathématique et de physique*. It consisted of short texts that were originally presented at meetings of the Académie and were collectively approved as worthy of publication (further quoted as *Mémoires de mathématique et de physique* 1702–1742). In each volume, between Fontenelle's *histoire* and the academicians' *mémoires* were inserted the eulogies (*éloges*) that Fontenelle had delivered at annual public meetings of the Académie when a member of the institution had died. In the *éloges*, Fontenelle presented the *persona* of a new type of scientist, explained the usefulness of the work by the deceased academicians, and placed their research in the more general context of the history of natural philosophy. Fontenelle's *éloges* quickly became popular, and from 1708 they were published independently.[6]

While the *mémoires* were strictly scholarly, in his *histoire* and *éloges* Fontenelle tried to translate scientific research into the categories and concepts of the educated readers of his time. Earlier research therefore suggested that Fontenelle was the first popularizer of science.[7] More recent research by French historians has indicated that it was too simplistic to label Fontenelle's activities as mere popularization. In fact, his work was much more sophisticated from literary and philosophical points of view, and individual historians such as Mazauric, Seguin, Claudine Poulouin, and Mitia Rioux-Beaulne have highlighted various aspects of it.

In the 1720s, Bignon ordered Fontenelle to prepare retrospective volumes describing the research at Académie before 1699. This project eventually resulted in 11 volumes, which were published under the title *Mémoires de l'Académie royale des sciences, depuis 1666 jusqu'à 1699* (published in 1729–1734). The first two volumes presented the history of the Académie between 1666 and 1699. The source for them was the registers of the Académie and the first volume of Du Hamel's history of the Académie, written in Latin (Du Hamel 1701). The author of the history of the early Académie (1666–1679) was Fontenelle; the second part (1680–1699) was authored by the astronomer Louis Godin (Mazauric 2007, 79).

These first two volumes were, however, somewhat unsystematically named *Histoire de l'Académie Royale des Sciences, depuis son établissement en 1666 jusqu'à 1699* (further quoted as Fontenelle and Godin 1733). The remaining volumes bear the aforementioned title *Mémoires de l'Académie royale des sciences* (sometimes labelled as *MARS*) and contain texts by academicians that were published or written in 1666–1699 (further quoted as *Mémoires de l'Académie royale des sciences* 1729–1734). For example, the third volume of *MARS* comprises of the works of Claude Perrault; the seventh volume contains reports and records of the many measurements taken by academicians on their travels, expeditions, and missions. The eighth volume contains the selected works of Cassini. The tenth volume includes smaller works and short texts by academicians from the early 1690s. This whole division of *HARS* and *MARS* volumes might be rather confusing.

[6]The *éloges* have been published in the volumes 6 and 7 of Fontenelle 1990–2001.

[7]Marsak 1959; Wade 1977; Shank 2018, 227–229; Gaukroger 2010b, 236–241.

The *HARS* volumes represent the research of the Académie from 1699 onwards, while the *MARS* volumes cover the history of the Académie up to 1699.[8] In addition, in the last third of the seventeenth century, academicians also published their papers, short tracts, and monographs under their own names. In the following chapters, I draw on all of these sources.

4.2 The Inexhaustibility of Nature

While Renaissance natural philosophy assumed that the world was closed and finished, optical instruments showed that this idea would have to be abandoned. Immediately after the invention of the telescope, Galileo realized that his short writing *Sidereus nuncius* (1610) had certainly not exhausted the potential of the hitherto hidden world: 'Perhaps more excellent things will be discovered in time, either by me or by others, with the help of a similar instrument...'[9] In a letter to Francesco Ingoli in 1624, Galileo wrote that there were many stars 'which are not observable with the telescopes built so far'.[10] In 1645, Rheita stated that recent discoveries concerning Saturn and Jupiter suggested that many things in the heavens 'were hidden from us' (Rheita 1645, 181b). In his work *De vero telescopii inventore*, Borel also asserted that with the improvement of telescopes, a better understanding of the distribution of the heavenly bodies would be obtained (Borel 1655, 50). The Magdeburg experimentalist Otto von Guericke and Huygens also expected that with better telescopes, natural philosophers would see more stars located at immense distances.[11] Gassendi even expressed concern about whether astronomers would ever be able to calculate the final amount of stars if they kept discovering more and more stars with ever-improving telescopes (Gassendi 1658, vol. 1, 505b).

In his *Micrographia* (1665), Hooke boasted about having discovered many more stars in the Pleiades with his long telescope than Galileo. According to Hooke, it was likely 'that the meliorating of Telescopes will afford as great variety of new Discoveries in the Heavens, as better Microscopes would among small terrestrial Bodies.'[12] In one of his letters, Leeuwenhoek wrote about the blood vessels of microorganisms, which were as tiny in relation to human blood vessels as human hair was tiny compared to the circumference of the globe. In the mysterious bowels of nature, Leeuwenhoek continued, there were wondrous dimensions: 'all we have as yet discovered is but a trifle in comparison to what lies hidden in the great treasure

[8] The overview of volumes of *HARS* provide Marchal 1997, 193–195; Mazauric 2007, 77–88; Seguin 2012; Seguin 2023.

[9] Galilei 1890–1909, vol. 3, 60, English translation in Galilei 1989a, 36.

[10] Galilei 1890–1909, vol. 6, 525, English translation in Galilei 1989b, 168.

[11] Guericke 1672, 43b; Huygens 1888–1950, vol. 19, 267.

[12] Hooke 1665, 242, similarly Boyle 1775, vol. 3, 51.

of nature.'[13] Following early modern natural philosophers and astronomers, Fontenelle also repeatedly emphasized the inexhaustible variety of nature and the infinite number of natural objects (Fontenelle and Godin 1733, vol. 1, 268).

I mention these quotations and paraphrases to demonstrate the widespread belief of natural philosophers and astronomers in the dependence of knowledge on observational technology. This relationship gave knowledge a temporal dynamic. Blumenberg noted that 'through the telescope, the contemplation of the heavens acquires a historical character: In a situation in which the cosmic horizon of experience had been constant since primeval times, the discovery of the telescope signifies a caesura, beyond which a continuous increase in the accessible reality could be anticipated.'[14] Ladina B. Lambert explained that while the telescope revealed phenomena invisible to the naked eye, 'it never showed enough and always implied more. The telescope thus drew attention to its limitation while simultaneously suggesting that the potential for new discoveries may be inexhaustible' (Lambert 2002, 36). The experience of using optical technology thus revealed three features of the new regime of knowledge, and all of them were linked to time: humans were gradually discovering parts of nature's inexhaustible riches, the future hid an unspecific amount of further discoveries, and the horizon of human experience was permanently shifting.

In his book on microscopic observation, Borel wrote generally that new things were constantly being discovered in nature (*nova semper in natura deteguntur*), new secrets were being revealed to human view, 'new gems are daily being drawn from the inexhaustible treasury of nature' (Borel 1656, fol. a2r). Other philosophers also used the metaphor of the treasury of nature[15] to indicate the dynamics of empirical research, which was carried out by humans over time gradually, one day after another. The discovery of the secrets of nature thus followed from a slow and methodical human activity, not from a fixed metaphysical order and statically understood limits of human knowledge. Natural philosophers coped with the inexhaustibility of nature through the purposeful reshaping and expansion of experience made possible by observational and experimental apparatus.

When we return to Hooke's 1692 lecture on telescopes and microscopes, we can notice another feature of the temporal regime of knowledge. Hooke sought to allay the doubts of some contemporaries who believed that optical instruments had exhausted their potential for discovery. Hooke insisted that further discoveries could not be doubted, 'if fit Methods and fit Instruments be apply'd, and prosecuted with Diligence' (Hooke 1726, 262). Philosophers would discover unimaginable things that resembled nothing in nature, 'certain non-pareils of nature' (Hooke 1726, 265). In his article on the concept of 'non-pareils', Hamou explained that early modern natural philosophers realized that, given the inexhaustibility and diversity of nature, they could expect to discover other objects that were completely unpredictable, for which they lacked a pattern and which were unknowable (Hamou 1999b, 16–17).

---

[13] Leeuwenhoek, letter to Huygens, 20 May 1679, in Leeuwenhoek 1939–1999, vol. 3, 66.

[14] Blumenberg 1996b, 420; English translation in Blumenberg 1985, 373.

[15] Glanvill 1668, 7; Sprat 1734, 383–388; Charleton 1654, 5; Hevelius 1673, vol. 1, 26.

Such beliefs, as shown by Hamou, can be found in Bacon, Galileo, and Gassendi.[16] In his *Préface sur l'utilité des mathématiques et de la physique*, Fontenelle also observed that the Ancients would have been greatly surprised by what their descendants had discovered with the aid of the telescope: an infinite number of objects that had been unknown to their predecessors. The Ancients had not even considered the existence of such objects (Fontenelle 1994, 42). From the context of Fontenelle's writings, we may reasonably infer that Fontenelle believed that such a surprise awaited the present generation as well.

In his speech, Hooke pointed out that the most famous example of such a 'non-pareil' object was Saturn's ring, discovered by Huygens in 1655. Huygens himself was aware of the unusualness of the object and therefore alerted his readers to the fact that the power and majesty of nature were infinite. Nature was constantly presenting with new patterns (*specimina*) and thus alerting humans to all that remained hidden within it (Huygens 1888–1950, vol. 15, 301).

Hamou explained that Saturn's ring and other 'non-pareils of nature' revealed two things: the excess of the visible beyond human knowledge and the limitations of human cognitive abilities. Both insights brought natural philosophers to believe that they had to change their mindset and adopt a 'disposition to expect the unexpected,' as Hamou aptly noted (Hamou 1999b, 20). In other words, natural philosophy was to avoid premature definiteness and was to learn not to rule out any possibilities in advance.

Empirical discoveries confirmed the rightfulness of this stance. In 1677, Cassini published a report on his discovery of various parts of Saturn's ring (today known as the Cassini division). Cassini explained that after Huygens discovered the planet's ring and its largest moon, he himself discovered two more satellites in 1671 and 1672, 'and then it seemed that nothing new could be discovered concerning the planet.' But Cassini admitted that he had to reconsider this view. His new discovery in 1677 proved that there was always something new to observe on Earth and in the sky (*Mémoires de l'Académie royale des sciences* 1729–1734, vol. 10, 582–583). Further investigations have confirmed this view. In 1684, Cassini discovered two more satellites of Saturn (a total of five were known at the time) (*Mémoires de l'Académie royale des sciences* 1729–1734, vol. 10, 694–704). This time he began his account of the discovery, which was partly reprinted in the *Transactions of the Royal Society*, differently. The variety of discoveries that had been made with telescopes, Cassini wrote, 'has incited Astronomers more strictly to Examine, if there were not yet something considerable, that had not been hitherto perceived'.[17] Thus, while in the early 1670s Cassini did not expect anything special when observing Saturn, by the 1680s, with more powerful telescopes, he was already deliberately searching for what astronomers had hitherto overlooked, assuming that there was still something to be discovered.

Unknown objects were also discovered in other areas, and Fontenelle and other members of the Académie regularly reported on them. The infinite variety and

[16]Bacon 1996–2008, vol. 11, 168–169; Galilei 1890–1909, vol. 7, 91; Gassendi 1658, vol. 1, 526a.

[17]*Mémoires de mathématique et de physique* 1702–1742, vol. 10, 694, English translation in Cassini 1686, 79.

creativity of nature were observed when examining an insect or an object such as a feather (Fontenelle 1702–1742, vol. 1, 39, 45). In relation to animal reproduction, Fontenelle wrote that the reproduction of earthworms might seem strange. But is not our idea of strange merely a manifestation of our ignorance? Do we know the limits of the variety with which nature has chosen to adorn her works? (Fontenelle 1702–1742, vol. 1, 41). When reporting on experiments with atmospheric pressure, Fontenelle emphasized that probable conclusions that rule out a possibility can be misleading. In physics, it sometimes happened that what seemed impossible turned out to be true—to the shame of reason (*à la honte du raisonnement*) (Fontenelle 1702–1742, vol. 2, 116). It turned out that in a contingent world, Hamou's 'disposition to expect the unexpected,' was always recommendable. The philosophers were to keep an open mind and not to rule out any possibility in advance based on speculative reasoning alone. The variety of nature transcended the reach of human reasoning and imagination.

The aforementioned examples demonstrate the growing conviction of natural philosophers that it was unwise to consider the possibility of concluding knowledge in the foreseeable future. In the mid-seventeenth century, Comenius hoped that with the approaching millennium it would be possible to summarize the content of the world in a single book like a thesaurus that would contain all that God wanted humans to know (Comenius 1964, vol. 1, 181–183). During the second half of the seventeenth century, it became clear that the production of a definitive and closed encyclopaedia that would grasp the whole world was not feasible. This traditional model of knowledge assumed that philosophers just drew more knowledge from a finite reservoir of objects: while the scope of human knowledge increased, the amount of the unknown decreased. The natural philosophers from the Académie rejected this indirect proportionality between known and unknown.

This brings us to the third feature of the temporal regime of knowledge brought about by optical instruments. Shortly after the discovery of the telescope, astronomers and philosophers believed that the new instrument was simply providing more findings. Later, they began to realize the unsettling relativistic effects of telescopic discoveries: the accumulation of knowledge also brought the cumulative production of the unknown. Hooke enthusiastically noted that 'every considerable improvement of Telescopes or Microscopes' presented 'new Worlds and *Terra-Incognita's* to our view' (Hooke 1665, fol. d2v). In *The Usefulness of Natural Philosophy* (1663), Boyle stated similarly, 'almost every day either discloses new Creatures, or makes new Discoveries of the usefulness of things; almost each of which hath yet a kinde of *Terra incognita*, or undetected part in it' (Boyle 1775, vol. 3, 230). Glanvill, in turn, spoke of an 'America of secrets, an unknown Peru of Nature' which opened to a new non-Aristotelian philosophy (Glanvill 1885, 153).

All these similes placed before human knowledge the vast mass of the unknown as a mysterious continent, as a static block. The static nature of the metaphor was somewhat misleading because it did not correspond to the dynamics of the new regime of knowledge, which always pushed the boundaries of the unknown further.

In his *Syntagma philosophicum*, Gassendi complained that humans did not have the sight that would enable them to glimpse the secrets of nature. Philosophers would improve the microscope and be able to see the composition of matter and the smallest particles that caused visible effects. Gassendi thus articulated his view that the boundary between the perceptible and the imperceptible was not fixed once and for all; the boundary moved thanks to ingenuity and skill of humans in the development of optical devices (Gassendi 1658, vol. 2, 560a).

Referring to this passage, Olivier Bloch explained that for Gassendi truth and ignorance were in a temporal, not dogmatic, relationship. The limits of human knowledge did not rigidly separate the known from the unknown; they continuously shifted. They were neither immutable nor essential, but movable and temporary (Bloch 1971, 102). Gassendi articulated this view when he divided the objects of knowledge into things manifest (*res manifestae*) and hidden (*occultae*), which were further divided into things inaccessible 'completely' (*penitus*), 'by nature' (*natura*), and 'temporarily' (*ad tempus occultae*).[18] Although some things remained unknowable in principle and completely, in many areas the boundaries of the known and unknown were fluid in time, rather than being dogmatically fixed.

Hence, according to Bloch, Gassendi's specific concept of scepticism. While acknowledging human ignorance in the knowledge of nature, Gassendi rejected the dogmatism of academic sceptics. According to the sceptics, the truth was inaccessible. Gassendi believed that the truth had not yet been discovered.[19] Gassendi thus alleviated scepticism by linking epistemological caution to time. Even though the expansion of knowledge might take a very long time, Gassendi understood his scepticism positively: the belief that the future would bring further discoveries prevented despair that could be caused by unknowability of nature (Bloch 1971, 103–104).

Gassendi's approach shows the birth of a new type of scepticism that eschewed the dogmatism of academic scepticism and the epistemological passivity of Pyrrhonian scepticism, which suspended judgment and opened up no further perspectives. Gassendi's mitigated scepticism, to borrow the well-known label of Richard Popkin (2003, 122–127), set up time against ignorance. Just as telescopes continually expanded the horizon of visibility for humans, knowledge as a whole was not condemned to remain within the absolutely understood boundaries of the knowable. Its extent changed and shifted. Therefore, no ignorance could be considered definitive, as it could always be interpreted as a phase in a long process.

The belief in the transience of empirical knowledge also came from the experience of using optical instruments. Practitioners of observational disciplines, such as Huygens and astronomers from the Observatoire, were increasingly aware of how much the extent of experience depended on both human optical theories and available technology.[20] So it turns out that humans themselves set the limits of empirical knowledge. As Ofer Gal and Raz Chen-Morris have pointed out, with the

[18]Gassendi 1658, vol. 1, 69a; cf. Detel 1978, 53–64; Hamou 1999a, vol. 1, 149–150.

[19]Bloch 1971, 100–101; Gassendi 1658, vol. 3, 412b–413a.

[20]Huygens 1888–1950, vol. 15, 231, 271, 295, cf. *Mémoires de l'Académie royale des sciences* 1729–1734, vol. 10, 83, 696–701; Fontenelle and Godin 1733, vol. 1, 416.

development of optical technology, scientific observation of the external world became an artificial procedure that relied not on the natural physiology of the human eye but on an understanding of the mathematical laws of optics. Optical instruments turned into the means of reason, which itself opened up other worlds inaccessible to natural human sensibility.[21] 'Being thoroughly mathematical, the telescope is not an extension of the eye but of reason' (Gal and Chen-Morris 2010b, 142).

The limits of visual experience thus became artificial and provisional because they depended on the sharpness of intellect, technological skill, and the inventive application of dioptrics. The expansion of knowledge did not mean a mere accumulation of new knowledge, but a deliberate expansion of the horizon of human experience. This was precisely the process in which the descriptive idea of progress was replaced by the prescriptive idea of progress.

4.3 *Hypotheses fingimus*

We have seen in previous chapters that the immense size and diversity of nature brought natural philosophers to believe that human knowledge covered only a small part of its extent. The second restraint of natural philosophy concerned the certainty and truth of knowledge. Mechanistic natural philosophy abandoned the ideal of *scientia*, i.e., demonstrative knowledge that sought to derive effects from causes (Serjeantson 2006), and replaced it with hypothetical knowledge. This process has been described from different perspectives by Larry Laudan, Desmond Clarke, Stephen Gaukroger, Sophie Roux, or Robert Pasnau, among others.[22] I want to demonstrate that the introduction of hypothetical methodology was part of the establishment of a temporal regime of knowledge.

The mechanistic programme in natural philosophy wanted to infer perceptible properties of objects from the properties of invisible particles such as shape, size, arrangement, and motion. The specific disposition of the corpuscles in a body was seen as the cause of its behaviour and visible properties.[23] The natural world, therefore, became a set of visible effects for which natural philosophers tried to find causal explanations in the form of assumed configurations of particles. These explanations left the perceptible world of Aristotelian qualities for the invisible microworld. As they could not be empirically verified, they were comprehended as hypothetical.[24]

[21] Gal and Chen-Morris 2010a; cf. Hamou 1999a, vol. 1, 277.

[22] Cassirer 1922, vol. 1, 478–483; Laudan 1966; Sakellariadis 1982; Clarke 1982; McMullin 2008, 2009; Roux 1998, Roux 2013a; Ariew 2011; Gaukroger 2010a; Pasnau 2017, 178–201.

[23] See Hooke 1665, fol. a2v; Boyle 1775, vol. 4, 68–78; Gassendi 1658, vol. 3, 22b.

[24] Clarke 1989, 134–135; Nadler 2000; 525–526; see also Joy 2006.

Typical representatives of this approach were the Cartesian natural philosophers Rohault and Régis. For both, physical explanations meant the discovery of the internal corpuscular structure of things that produced visible effects. According to them, natural philosophy sought to create a micromechanical model of perceptible phenomena. This model was always hypothetical because it only assumed a configuration of particles with certain properties that produced the perceptible effects.[25]

Representatives of mechanistic natural philosophy had to solve the problem of the equivalence of different hypotheses with respect to the same phenomenon (today called the problem of underdetermination) (Anstey 2019, 48). The mechanists often tried to explain this methodological difficulty using an analogy to clocks,[26] which was aptly formulated by Roger Cotes in his introduction to the second edition of Newton's *Principia* (1713): 'In mechanical clocks, one and the same motion of the hour hand can arise from the action of suspended weight or an internal spring' (Cotes 2004, 52).

The mechanists solved this problem in various ways. Rohault summarized his methodology as follows: 'Thus we must content for our selves for the most part, to find out how Things may be; without pretending to come to a certain Knowledge and Determination of what they really are; for there may possibly be different Causes capable of producing the same Effect, which we have no Means of explaining'.[27]

Although Rohault's words gave the impression of epistemological pessimism and methodological instrumentalism, he and other Cartesians denied that they were merely 'saving phenomena' by 'arbitrary hypotheses' (Régis 1690, vol. 1, 276). French Cartesians followed their master in attempting to create a physical system based on a limited set of metaphysical or physical principles. The Cartesians then insisted, with strong realist pretensions, that their hypotheses met certain criteria, such as simplicity, consistency with the metaphysical foundations of knowledge, systematic coherence with other hypotheses, explanatory scope, and agreement with empirical evidence. Hypotheses successfully meeting these criteria were convincing to Cartesians.[28]

The natural philosophers from the Royal Society implemented their own versions of mechanistic programmes, which in part coincided with Cartesianism. As Wilson noted, the imaginary models of Cartesians 'differed little from those favoured in mainstream Royal Society experimental philosophy' (Wilson 2019, 173). Besides similar ontological assumptions, both groups shared several methodological tenets. For instance, Boyle presented similar criteria for choosing hypotheses as those found

[25] Rohault 1671, vol. 1, 19–20, vol. 2, 4; Régis 1690, vol. 1, 274–275.

[26] Descartes 1996, vol. 8, 328; Rohault 1671, vol. 1, 21; Régis 1690, vol. 1, 275; cf. Mariotte 1678, 13, 133.

[27] Rohault 1671, vol. 1, 21; English translation in Rohault and Clarke, vol. 1, 14; cf. Régis 1690, vol. 1, 275. This aspect of Cartesian methodology was explained by Clarke 1989, 190 and 2018, 101–110; cf. McClaughlin 1977; Roux 2013b; Dobre 2013.

[28] Rohault 1671, 1, 21; Régis 1690, vol. 1, 276–278; cf. Clarke 1989, 228–231; Dobre 2019.

in the Cartesians.[29] The important difference between the British and Cartesian versions of mechanistic natural philosophy was that the British natural philosophers had no ambition to create a closed system out of individual micromechanical models, nor did they require a robust realist claim for their hypotheses. They wanted to avoid dogmatism, which they saw as too similar to the religious zeal that had been the cause of social disruption in the past. Rather, they followed Bacon's project of the instauration of the sciences emphasizing epistemic modesty and creating natural histories to serve as a basis for later inductions.[30] Time thus became an important part of the mechanistic methodology: the final judgement was postponed to the future.

When the Académie was founded, Cartesians were not accepted at all, because they were suspected of dogmatism and dependence on the authority of Descartes.[31] The representatives of the early Académie more likely shared epistemological ideals with British natural philosophy which could be the result of Gassendi's influence on both groups. As early as 1666, Huygens described to Colbert the methodological orientation of the new institution. He stated that the forthcoming society should 'work on natural history following the intention (*dessein*) of Verulamius'. This natural history was to consist of experiments and reflections. Huygens judged that such a programme represented the only way 'to arrive at a knowledge of the causes of all that can be seen in nature'.[32] Similarly, other representatives of the early Académie wanted the institution to pursue a Baconian programme of data collection in natural history rather than to develop a system in the Cartesian vein.

Therefore it is not surprising that in *the MARS* and *HARS* volumes academicians referred to the results of their work and the views of other philosophers as hypotheses.[33] However, the terminology was not settled, so in addition to *hypothèse*, the terms *conjectures, suppositions*, and *systèmes* were used.[34]

In his *Essay de logique* (1678), academician Edme Mariotte openly admitted that, given the vastness of nature and the limitations of human cognitive faculties, human knowledge had to be limited to the description of effects. Mariotte considered the speculative inference of causes from principles to be a mistake. The basis of physics was an inductive generalization from individual observations and experiments (Mariotte 1678, 133, 143, 145, 195). According to Mariotte, the natural philosopher had to be content with a comprehensive natural history of the effects of nature that had been experimentally observed. In his view, an hour's experiment often taught

---

[29] Boyle formulated such criteria in his unpublished treatise *The Requisites of a Good Hypothesis* (c. 1655); see Westfall 1956, 116–117; cf. Anstey 2019.

[30] Cf. Dear 1985, Dear 1995, 224–242, Dear 2001, 138–145; Sacco 2020, 17–40.

[31] See Fontenelle and Godin 1733, vol. 2, 379–381; Petit 1665, 234, 240, 253; Waard 1925; Roux 2017.

[32] Huygens, letter to Colbert, 1666, Huygens 1888–1950, vol. 6, 95, cf. similarly vol. 19, 268.

[33] *Mémoires de mathematique et de physique* 1702–1742, vol. 10, 527–528, 537, 572–574, 581–582; 37; cf. Fontenelle 1702–1742, vol. 1, 61, 116; vol. 2, 46, 50, 152.

[34] Mariotte 1678; 1699: 60; Fontenelle 1702–1742, vol. 2, 7, 29; Régis 1690, vol. 1, 276; cf. Roux 2011, 141.

much more than reasoning that takes years. Even if in this way the philosopher did not recognize all the causes, it was better than wasting time making speculative hypotheses that hindered 'the progress of physics.'[35]

In his four-volume *Essais de physique* (1680), Claude Perrault insisted that empirical knowledge of nature would never achieve certainty. He then referred to his own investigations as mere *essais*, because in physics one could only essay (*essayer*), seek, and formulate uncertain and probable hypotheses (Perrault 1680, vol. 1, unpaginated preface; vol. 3, 5–6).[36]

Representatives of the empirical disciplines also commented on the uncertainty of hypotheses and the problem of their equivalence in relation to the same phenomenon. Fontenelle was clearly aware of the problem: 'we can imagine figures which will satisfy the Phenomenon and which will not be however those that Nature employs there' (Fontenelle and Godin 1733, vol. 1, 94). In his *Mémoires pour servir a l'histoire des plantes* (1679), natural historian Denis Dodart stated that multiple causes could always be assumed for observed phenomena. Therefore, in the book, he and his colleagues formulated only a few conjectures concerning the nature and effects of plants, based on accumulated experience (Dodart 1679, 70, 295). Although Dodart would have liked more certainty for his conclusions, he could provide only 'reasonable conjectures' (Dodart 1679, 75, 103–105).

The academicians quoted above understood the problem of the equivalence of hypotheses as part of the temporal regime of knowledge. This meant that they considered the hypotheses tentative and provisional and believed that the future would show which one was correct. Instead of devising methodological criteria for selecting hypotheses, they recognized time as the decisive criterion. Representatives of the empirical sciences therefore did not understand hypotheses as absolute statements independent of time (as, for example, Descartes did). For them, hypotheses were always situated in time because they embodied the dependence of empirical knowledge on the set of data available at the moment.

Hypothetical reasoning resulted from the invisibility of corpuscles but theory of matter was not the only reason. Hypotheses were also introduced as methodological instruments coping with the inexhaustibility of nature, which threatened every inductively drawn conclusion. In a contingent world, it was hard to rule out the possibility that the future would bring findings that challenged current conclusions.

British natural philosophers addressed this problem by emphasizing the provisional nature of their hypotheses, or conclusions. Although Newton famously deprecated the framing of hypotheses, he formulated hypotheses, for example in the *queries* at the end of his *Opticks* (see e.g. Newton 2004, 139). According to Kirsten Walsh, in Newton's methodology, 'hypotheses provide a temporary platform from which further experimental work and/or theorising can be carried out' (Walsh 2019, 126). Describing Hooke's methodology, Hamou also stressed that the hypothesis had an operational character for the British philosopher. Its purpose was

[35] Mariotte 1678, 142–143; cf. Roux 2011, 82.

[36] Perrault 1680, vol. 1, unpaginated preface; similarly vol. 3, 5.

to serve as a basis for focalizing and accelerating the inductive process. Its typical feature was provisionality. The hypothesis was intended to orient experimental and observational research in a certain direction but could be abandoned and replaced after partial results were achieved (Hamou 1999b, 24).

The natural philosophers and natural historians of the Académie also used hypotheses as temporary starting points that helped to guide experimental research as further knowledge came. In their view, hypotheses could be revoked or modified without a loss of confidence in the research itself. For instance, Dodart wrote that the properties of plants were known only in the form of general assumptions based on inductions to which something can always be added. Hence no certain conclusions could ever be formulated (Dodart 1679, 75). In the collective work *Mémoires pour servir à l'histoire naturelle des animaux* (1671), the editor, Claude Perrault, wrote that the conjectures they included among the results of their observations were only preliminary. The authors were prepared to retract them if contradictory observations appeared (Perrault 1671, fols. i-r, e-v). In his *Histoire des plantes* (1698), the botanist Pitton de Tournefort also asserted that he proposed inductions drawn from his observations as mere conjectures, which he was willing to correct if necessary (Tournefort 1698, unpaginated preface). Similar views were formulated by Claude Perrault (1680, vol. 1, unpaginated preface) and Mariotte (1678, 33).

In the *HARS* volumes, Fontenelle repeatedly reminds his readers of the temporality of hypotheses in empirical disciplines. Nature was too vast and changeable for philosophers to believe in the permanence of their findings: 'We must always look back, always retrace our steps, and never believe ourselves to be in full possession of physical truths' (Fontenelle 1702–1742, vol. 3, 62; cf. vol. 4, 52; vol. 5, 16; vol. 8, 15–16). In another year, Fontenelle wrote about astronomy that its findings were always subject to revision, and the work in the field never ended. Astronomy was always in motion, like the stars it described (Fontenelle 1702–1742, vol. 8, 96, 120).

Hypothetical reasoning understandably lacked the necessity of demonstrative knowledge. Thus, natural philosophy in the second half of the seventeenth century definitively broke with the ideal of *scientia*. In a teleological, closed, and finite cosmos that corresponded to the human sensory equipment, such an ambition made sense. In a contingent, diverse, and unclosed universe, such an ideal seemed unattainable.

Locke saw the inability of natural philosophy to recognize the causes of things as a manifestation of the limited capacities of depraved human beings.[37] The natural philosophers of the Académie did not comprehend hypothetical knowledge as the bitter fate assigned to human beings in the history of salvation. They did not count at all on the possibility of achieving a certain demonstrative knowledge in the field of empirical research. Dodart argued that anyone who wanted to publish only what was certain or perfect in physics (*physique*) would publish nothing (Dodart 1679, 104–105, 327–328). Therefore, two new norms of the reliability of knowledge

[37] Locke 1975, 640; cf. Osler 1970, Specht 2009, Rogers 2010.

took the place of the certainty of demonstrative knowledge: probability and temporary hypotheses based on limited amount of data.

While the concept of probability was born in the mathematical conceptualization of games and betting (Daston 1998; Hacking 2006), some philosophers from the Académie introduced the concept into natural philosophy. Huygens famously subscribed to probability at the beginning of *Traité de la Lumière* (1690) and in his *Cosmotheoros* (Huygens 1688–1950, vol. 19, 455, vol. 21, 687–689). Perrault and Mariotte took a similar stance even earlier.[38] Fontenelle separated geometry from physics precisely with respect to certainty of knowledge. While certain demonstrations were possible in geometry, physical research did not carry a certainty of conclusion, even when mixed with mathematics, and therefore its conclusions were only probable (Fontenelle and Godin 1733, vol. 1, 234, 253, cf. Fontenelle 1702–1742, vol. 1, 72). Probability also appeared in mechanics: In his treatise on gravitation, Varignon wrote that he was only providing "probable conjectures" (Varignon 1690, unpaginated *avertissement*). The notion of probability did not apply only to mathematical disciplines. Even the anatomist Alexis Littre referred to his reasoning as probable when describing his anatomical investigations (*Mémoires de mathématique et de physique* 1702–1742, vol. 3, 93–94, 112).

Probable and temporary hypotheses compensated for their lower degree of certainty by giving research more freedom and flexibility, at least in the eyes of some academicians. They felt that hypotheses freed natural philosophy from the dogmatism of following unchanging principles because they opened up the possibility for the creativity of reason and inclined the mind to 'expect the unexpected'. In his *Cosmotheoros*, Huygens wrote that the uncertainty of knowledge applied to all of physics, in which at most probable knowledge could be formulated. Huygens did not seem to perceive this as a failure or anthropological deficiency. He emphasized that the allure of science lay not in truth and certainty, but in the discovery of the unknown and the hidden (Huygens 1888–1950, vol. 21, 689). The ideal of demonstrative knowledge no longer seemed important to Huygens. What mattered to him was the further advancement of knowledge. The essence of empirical natural philosophy was the continuous expansion of human knowledge in time, not the completion of knowledge in indisputable certainty.

In the introduction to his *Essais*, Claude Perrault also pointed out the freedom to frame hypotheses. Hypotheses were supposed to correspond to facts but otherwise Perrault advocated unbridled imagination in forming them—'the most bizarre imaginations are allowed'. According to Perrault, empirical research consisted of piling up various hypothetical explanations, 'none of which will ever be true'. Like Huygens, Perrault stressed that it was not evidence and certainty that were the source of the 'beauty and excellence' of human knowledge, but freedom of thought (Perrault 1680, vol. 1, unpaginated preface).

---

[38] Perrault 1680, vol. 1, unpaginated preface and 2; vol. 3, 6; Mariotte 1678, 23, 33, 197; cf. Roux 2011, 99.

Perrault developed this idea further in the third volume of his *Essais*. Here he wrote that the beauty of philosophy lay in diversity. It was better to have several probable theories with different degrees of probability than a single most probable one. Only a diversity of different systems could satisfy the natural human desire for knowledge. Every theory explained some phenomena, and when a phenomenon eluded them, philosophers invented another system. Perrault asserted that as long as the world lasted, theories would alternate in response to the abundance of phenomena. Philosophers would be made to create other theories, 'without ever hoping to discover the true one' (Perrault 1680, vol. 3, 6).

For Descartes, the hypothesis was only a provisional solution to serve practical ends, and it would one day be replaced by definitive knowledge.[39] For the representatives of the empirical research conducted at the Académie, the hypothesis represented a form of natural philosophical knowledge that was adequate to the contingency of the human condition. As everyday practitioners of empirical natural philosophy, they preferred hypotheses as methodological instruments that had the advantage of avoiding definite statements. Hypotheses thus provided their investigations with freedom that allowed continuous modifications resulting from incoming observational and experimental data.

Natural philosophers of the Académie thus used hypotheses as an open-ended form of knowledge that was deliberately formulated without claim to demonstrative certainty; as temporary, provisional, and incomplete. The hypotheses embodied a collection of knowledge that changed dynamically over time and that provided its authors with two advantages: epistemic independence of the supposedly absolute principles and flexibility concerning incoming data. The goal of knowledge was no longer certainty, but gradual and free improvement, which was understood as a slow liberation from error rather than a movement towards truth. Therefore, the hypothetical methodology must be counted among the conditions making the prescriptive idea of scientific progress possible.

4.4 Sage Pyrrhonism

The natural philosophers of the Académie comprehended the hypothetical form of knowledge not as pessimistic agnosticism, but as a reasonable epistemic stance that protected natural philosophy from dogmatism. The academicians distinguished two forms of dogmatism. As for the depth and extent of knowledge, natural philosophers feared excessive epistemic claims striving to explain the causes and essences of phenomena. As for the structure of knowledge, academicians rejected binding knowledge into one dogmatic system because that would lead to sterility and intellectual slavery.

[39] Descartes 1996, vol. 9–2, 323; cf. Ariew 2011.

Trying to avoid both forms of dogmatism, the natural philosophers of the Académie took a position that Fontenelle labelled 'sage pyrrhonism' (*un sage pyrrhonisme*) (Fontenelle 1994, 49). Fontenelle seemed to use this term deliberately. He apparently wanted to avoid the notion of scepticism that was associated with academic scepticism, which insisted on the unknowability of things. In the history of philosophy, Pyrrhonian scepticism represented a more sophisticated position that, instead of dogmatic agnosticism, presupposed an equipollence of judgments (*isosthenia*) resulting in a suspension of judgment (*epochē*). In his *Préface sur l'utilité des mathématiques et de la physique* Fontenelle clearly explained this pyrrhonism[40]:

> [Unlike the ancient philosophers] at present the Académie of Sciences does not set about Nature but in little parcels: No general system for fear of falling into those rash ones, which do but too well accommodate the impatience of our Minds, and which being once established do oppose the Truths that present themselves to our View. To Day we make sure of one Fact, to Morrow of another that has no Relation to the former. We venture sometimes to guess at Causes, but then we produce 'em for no more than Conjectures. Thus the Collection which the Académie offers to the Publick, is composed of Pieces that are loose and independent of each other, of which each private Man, who is the Author is responsible for the Facts and Experiments, and of which the Society does not allow the Conclusions, but with all the restriction of a discreet Scepticism (*un sage pyrrhonisme*).

Fontenelle's words indicated that sage pyrrhonism was a set of several methodological tenets: collecting isolated data over a long period, avoiding hasty generalizations and system-building, proceeding slowly and incrementally by making only conjectures, and refraining from judgment when empirical data was lacking. In this section, I want to explain how these tenets gave the research at the Académie temporal dynamics. Generally, it can be said that this sort of scepticism was meant above all as a gradual move from errors and prejudices to the true knowledge in the unspecified and distant future.

The roots of the sceptical position in natural philosophy went back to Francis Bacon. In proposing his project of a great instauration of sciences, Bacon insisted that natural philosophy that truly wanted to serve humanity should rid itself of anthropocentric prejudices, dogmatism, and the worship of authority. Furthermore, philosophers should not succumb to a false sense of complacency because such a conviction brought theoretical stagnation: 'opinion of plenty is among the greatest causes of poverty' (Bacon 2004, 10). In the third phase of Bacon's project, called *historia naturalis & experimentalis*, the philosophers were meant to collect empirical data as independent pieces of evidence without rushing to a final form of philosophy: 'I nevertheless wait for the ripened crop, and do not try to mow the moss or green corn' (Bacon 2004, 38, 177). With this metaphor, Bacon wanted to articulate his concern about the rush of speculative reason, which he saw as a tendency to formulate conclusions too quickly, resulting in the construction of false systems. On the contrary, knowledge needed to be slowed down and spread out over time and across generations (Bacon 2004, 24).

---

[40] Fontenelle 1994, 49; English translation in Fontenelle 1727, xvi–xvii.

Bacon's emphasis on slowness was later championed by representatives of the Royal Society. Its advocates presented the Society as 'the bank', 'the common stock', 'the public treasury', or 'the public registers' in which data was collected to be passed on to future generations. This data was collected into 'unfinished histories', not closed systems (Sprat 1734, 64, 74–75, 115). Boyle summarized this attitude in his *Defence* (1662):

> It was not my chief design to establish theories and principles, but to devise experiments, and to enrich the history of nature with observations faithfully made and delivered; that by these and the like contributions made by others, men may in time be furnished with a sufficient stock of experiments, to ground hypotheses and theories on (Boyle 1775, vol. 1, 212).

According to Thomas Sprat, the ancient philosophers were wrong in being too quick and hasty to determine the causes of things without a sufficient amount of experience. Sprat complained in a Baconian vein that ancient authors should have produced only histories of nature, not conclusions: 'if they had only set Things in a way of propagating Experiences down to us, and not impos'd their Imaginations on us, as the only Truths.'[41] Sprat noted that, unlike the Greeks, members of the Royal Society merely collected and sorted facts without forming theories and passed them on to the future. 'By this, they have made a firm Confederacy, between their own present Labours, and the Industry of future Ages [. . .] if it had been begun in the Time of the Greeks [. . .] what Depth of Nature could by the Time have been hidden from our View?' (Sprat 1734, 116; cf. Hooke 1705, 21–24).

When Gassendi presented the biographies and research of Renaissance astronomers, he also emphasized the continuous collecting of data (Gassendi 1658, vol. 5, 365–386). Observation, according to Gassendi, was the basis of astronomy, not theoretical reasoning (Gassendi 1658, vol. 6, 46b). While reason could repeatedly fall into error, observational data was unique and irreplaceable (Gassendi 1658, vol. 6, 29b, 209a, 504). Unlike theories and hypotheses, data retained their value permanently and could serve posterity in the future.[42] Like the British followers of Bacon, Gassendi also subordinated knowledge to a temporal regime. The collection of data was intended primarily for later use, not for prematurely devising theoretical solutions.

Many members of the Académie followed Bacon and Gassendi in this point. The academicians conducting the experimental and observational research focused on the acquisition of facts and their collection, storage, and comparison (Hahn 1971, 31–35). They regarded the framing of explanatory hypotheses as a secondary task that they did not always fulfil. Dodart aptly wrote that the Académie was founded for doing experiments, not reasoning (Dodart 1679, 314–315). In the *HARS* volume for the academic year 1701, Fontenelle wrote—similarly to Sprat—that some observations did not fit any theory at the moment, although they were very accurate. They were therefore stored in the treasury (*trésor*) of the Académie, from where they could be easily retrieved (Fontenelle 1702–1742, vol. 3, 56). In the *Mémoires pour servir à*

[41] Sprat 1734, 117; cf. Hooke 1665, 87; Hooke 1705, 3–4.

[42] Gassendi 1658, vol. 6, 474b; Van Helden 1976.

l'histoire naturelle des animaux, Claude Perrault wrote that he and his colleagues gathered their observations into *mémoires*, which served as storehouses (*magasins*) from which one could take things when one needed them (Perrault 1671, fol. i-r). Fontenelle also stated that the volumes of the Académie were intended to preserve the acquired knowledge for posterity during historical upheavals (*révolutions*). Had the Ancients had a similar means, none of their knowledge would have been lost (Fontenelle 1702–1742, vol. 1, 118).

In presenting their results, both observers of astronomical phenomena and experimenters in laboratories preferred descriptive language, wanting to avoid any theoretical apriorism. Academicians wanted to elicit in their readers the impression that they were presenting mere data and provisional hypotheses with no claim to absolute validity.

In the preface to *Mémoires pour servir à l'histoire naturelle des animaux*, Claude Perrault emphasized that the natural historians were only describing what they themselves saw 'with simplicity and without ornament'. In their text, they wanted to present natural objects 'as in a mirror' that inserted nothing of itself into the resulting picture. Although they added some reflections to their observations, they hoped that readers would not dwell on them too much and would concentrate on the facts. Facts were also the only means by which the natural historians wished to point out the errors of the great figures who had written before them (Perrault 1671, fol. i-r).

The *Mémoires pour servir à l'histoire naturelle des animaux* indicated that natural historians and philosophers often proceeded by publishing observed phenomena and measured values without formulating a specific theoretical framework to explain them. Facts were considered crucial, having value independent of time. In the 1670s and 1680s, Huygens, Mariotte, and Cassini wrote various texts in which they described their research. These documents often took the form of a descriptive summary of various experiments and observations. In some cases, the authors derived a hypothesis from data; in others, they settled for merely presenting an overview of facts (*Mémoires de l'Académie royale des sciences* 1729–1734, vol. 10, 526–528, 529–536, 572–575, 593). Claude Perrault followed the same line of reasoning: 'It is much safer to entrench (*se retrancher*) oneself in the certainty of simple facts than to expose oneself to the peril of misunderstanding the consequences one wants to draw' (Perrault 1680, vol. 3, 14).

Astronomers regularly presented this epistemic practice in their texts. Most astronomical papers published as part of the *MARS* were lists of observations of singular events such as eclipses and descriptions of the circumstances under which the observations occurred (*Mémoires de l'Académie royale des sciences* 1729–1734, vol. 7, 211–233, 244–270). Astronomers deliberately avoided theoretical considerations or preferences for a cosmological system that would serve as the interpretative framework for observational data. They consistently spoke only of hypotheses, opinions, and systems of which none were recognized as valid. Indeed, as late as the end of the seventeenth century, they were still referring to heliocentrism as a hypothesis, or one system among others.[43]

The astronomers from the Observatoire clearly favoured long-term data collection at the expense of creating new theoretical frameworks.[44] They left cosmological problems, such as gravity in tourbillons, to physicists (Fontenelle and Godin 1733, vol. 1, 40, 100–101). This is also why astronomers did not appreciate so many advances in astronomical theory. When Cassini described the history of modern astronomy, his heroes were not the theorists Copernicus and Kepler, but the systematic observer Tycho (*Mémoires de l'Académie royale des sciences* 1729–1734, vol. 8, 29). In astronomy, the gathering of observations was decisive, not conjecturing and speculating. Similarly to Dodart, Cassini would probably also say that the Observatoire was built for observing, not for reasoning.

Another characteristic feature of sage pyrrhonism was the opposition to drawing hasty conclusions and to the premature creation of systems. Royal Society advocates also held this position, but they were primarily opposed to Greek science and the system of Aristotelian natural philosophy.[45] Members of the Académie saw the threat of dogmatism more in Cartesian natural philosophy. In his *éloge* for Régis, Fontenelle did not hesitate to question Régis's Cartesian philosophical system for being dogmatic and suppressing the diversity of nature in the name of the coherence and unity of the system (Fontenelle 1990–2001, vol. 6, 147). Fontenelle considered Cartesian system of nature to be a manifestation of a hasty dogmatism that tried to squeeze natural phenomena into the narrow confines of a single philosophical theory and prevented the free framing of hypotheses. In the *HARS* volumes, Fontenelle often warned philosophers against subordinating nature to what the limited reach of their reason and imagination told them. Philosophers had no right to prescribe their rules to nature, and if they wanted to know it, they had to start with it—not with their ideas (Fontenelle 1702–1742, vol. 4, 76, cf. vol. 7, 27, vol. 9, 97).

These views also motivated the academicians' ambivalence towards Descartes. Cassini, reflecting back on the modern history of astronomy, admitted that Descartes contributed to the improvement of astronomy in his *Principia philosophiae* (1644). At the same time, Cassini immediately downplayed Descartes's importance: 'he was more interested in reasoning than in observing' (*Mémoires de l'Académie royale des sciences* 1729–1734, vol. 8, 30).

Huygens was even harsher in a letter to Pierre Bayle from 1693. According to Huygens, Descartes was successful in passing off his conjectures and imaginations as truths. In Huygens's view, Descartes should have presented his system as a conjecture that was to draw probable conclusions from principles. These conclusions should have been further explored by other *bons esprits*. Instead, according to

[43]Even at the end of the seventeenth century, the Copernican theory remained a hypothesis among others for Cassini, see *Mémoires de l'Académie royale des sciences* 1729–1734, vol. 10, 98, 596, 645, cf. Fontenelle and Godin 1733, vol. 2, 100.

[44]Fontenelle and Godin 1733, vol. 1, 40, 100–101.

[45]Glanvill 1885, 125–163, 182–183, 196; Glanvill 1668, 89–92; Sprat 1734, 97–110.

Huygens, Descartes believed that he had found the truth, and thus committed great prejudice against the progress of philosophy (*grand prejudice au progrès de la philosophie*). Huygens further complained that Descartes's followers thought they knew the causes of everything and often wasted their time defending their master's doctrine instead of investigating the true causes of natural phenomena, about which Descartes left only his chimeras.[46] In Huygens's view, systematical philosophy of Descartes and his followers lacked precisely the key features of the temporal regime of knowledge: scepticism about knowing causes and substances of things, caution, probability, temporality, openness to new views and revisions, and long-term empirical investigation of nature.

The sceptical resistance to prematurely formulated systems also came from the experience of the constant increase of new knowledge. Fontenelle, Dodart, Cassini, and Philippe de La Hire all agreed on the constant flow of discoveries. They stated in exaggeration that philosophers discovered something new every day.[47] The consciousness of a permanent increment of methodically generated knowledge naturally gave gathered knowledge the stamp of provisionality. Therefore, natural philosophers required methodological prudence in the evaluation of knowledge. In his *Entretiens*, Fontenelle argued in connection with the plausibility of the hypothesis of the existence of life on the moon: 'You're going pretty quickly [. . .] You should never give more than half your mind to beliefs of this sort, and keep the other half free so that the contrary can be admitted if it's necessary.'[48]

In *Mémoires pour servir à l'histoire des plantes,* Dodart also stressed that he and his colleagues did not want to embark on hasty assertions and left a lot of room for vague ideas and open-ended theories. Natural historians did not want to decide what was possible and impossible by formulating binding hypotheses. They wanted to concentrate on the description and results of their experiments without dismissing as false what they failed to replicate in their experiments. Such failures could have unexpected causes, such as differences in climate, and therefore did not immediately imply a rejection of the experimental results.[49]

Sage pyrrhonism thus required an open mind that would develop the 'disposition to expect the unexpected' based on experience with unpredictable natural phenomena. Such a research would not stagnate on the principles of a single system that wanted to explain everything. Instead, a sage mind was open to revisions, modifications, and additions as research expanded.

Another important tenet of sage pyrrhonism was the suspension of judgment. This attitude was simple and pragmatic. If academicians thought that they did not

[46] Huygens, letter to Pierre Bayle, 26 February 1693, Huygens 1888–1950, vol. 10, 404–406.

[47] Dodart 1679, 6; Fontenelle 1994, 47; Fontenelle and Godin 1733, 45; *Mémoires de l'Académie royale des sciences* 1729–1734, vol. 10, 703; vol. 9, 3; *Mémoires de mathématique et de physique* 1702–1742, vol. 2, 71.

[48] Fontenelle 1990–2001, vol. 2, 57, English translation in Fontenelle 1990, 37.

[49] Dodart 1679, 42–43; *Mémoires de mathématique et de physique* 1702–1742, vol. 5, 134, vol. 7, 32.

have enough data to answer a question reliably, they left it deliberately unanswered and thus open to further investigation. Similarly to other methodological features of sage pyrrhonism, the suspension of belief postponed the attainment of a solution out of concern about hasty reasoning. Many examples of this position appeared in the publications of the Académie.

In 1668, the academicians Duclos, Perrault, and Mariotte argued about the circulation of sap in plants. Fontenelle articulated the opinion of the collective body of the Académie on the dispute in a single sentence: 'The greatest part of wisdom consists in the suspension of judgment' (Fontenelle and Godin 1733, vol. 1, 63).

Another example from Fontenelle's history of the Académie dates from 1679. Fontenelle reported on Mariotte's investigation of light, pointing out that Mariotte failed to give the true causes of the observed behaviour of the rays. According to Fontenelle, Mariotte, therefore, confined himself to merely stating his observations, and otherwise shut himself up in 'wise ignorance' (*une sage ignorance*), which was always better than presumptuous hypotheses (Fontenelle and Godin 1733, vol. 1, 295).

In 1700, academicians investigated the behaviour of animal muscles during vomiting. In this context, Fontenelle noted that the findings obtained by observing one species of animal must not be transferred without justification to other species: 'There is still nothing safer than to suspend the general propositions' (Fontenelle 1702–1742, vol. 2, 29). In the *HARS* volume for the academic year 1702, Fontenelle described a new method of amputation the success of which was not clear. Fontenelle therefore concluded his report by writing that the Académie suspended its judgment and awaited more experience (Fontenelle 1702–1742, vol. 4, 34). In 1705, academics argued over the most appropriate barometer design. Fontenelle concluded his description of the debate by stating again that the Academy was postponing its decision and would wait for further experiments. According to Fontenelle, the Académie did not only want to show the public its discoveries, but also its doubts, and would be very pleased if its doubts helped someone else to make new discoveries (Fontenelle 1702–1742, vol. 7, 21).

In the same *HARS* volume of, Fontenelle described the experiments carried out by Johann Bernoulli with mercury in a barometer. Bernoulli came to conclusions that did not coincide with the observations and findings of the members of the Académie. In Fontenelle's narrative, the personified institution again decided to 'suspend its judgment'. Fontenelle continued by saying that since philosophers explored nature a little bit, they also knew how dangerous it was to make hasty judgments about natural effects. The next year, therefore, there would be further tests of Bernoulli's conclusions. With his usual aptness, Fontenelle formulated the academicians' final opinion: 'Such a new idea could not be accepted without much discussion, nor could such a beautiful idea be rejected without the same discussion.'[50]

---

[50] *Mémoires de mathématique et de physique* 1702–1742, vol. 2, 8. See other examples in Seguin 2006.

The natural philosophers of the Académie often explained the suspension of judgment simply as the result of a lack of data. When Wilhelm Homberg conducted experiments with an air pump, he drew some conclusions but found them unconvincing. He insisted that they would have to be verified by new experiments (*Mémoires de l'Académie royale des sciences* 1729–1734, vol. 10, 280–283). Similarly, in anatomical research, the anatomists refrained from reaching a definitive conclusion, insisting that they needed to gather more experience (*Mémoires de l'Académie royale des sciences* 1729–1734, vol. 10, 466–467). In 1690, the academicians were cautious about the flattening of Earth that Newton proposed. While they acknowledged that the pendulum experiments at the equator suggested a flattening of Earth, the differences in the length of the pendulum in Paris and equatorial Cayenne may have been due to other factors such as temperature and humidity. They therefore left the question undecided (Fontenelle and Godin 1733, vol. 2, 102).

In *Mémoires pour servir a l'histoire des plantes,* Dodart stated that natural historians had not yet gathered enough experience to draw conclusions (Dodart 1679, 187). The word 'yet' was important because it indicated the academicians' belief that the lack of data was only a momentary state that would change in the future. The presented state of knowledge was therefore only a captured moment between the long past of natural history and its future.

Suspension of judgment motivated by lack of data could also be found in the astronomical research of Cassini. In 1667, Cassini insisted that he dared not yet make any definite claims about the rotation of Venus because he had not made enough observations of spots on its surface (*Mémoires de l'Académie royale des sciences* 1729–1734, vol. 10, 471; cf. Cassini 1810, 308). Cassini showed the same caution in his treatise on the comet that appeared in the sky in 1680. In his view, the nature of comets was still very much hidden (*fort cachée*) and all astronomers could do was compare reliable observations of comets by noting what comets had in common and what was special (Cassini 1680, 7). In 1699, Cassini still felt that he did not yet have enough observations to form a sufficient 'induction' that would allow him to confirm his theory of the return of comets: that is, to confirm the identity of emerging comets and to predict their return. At this point, it was only possible to collect observations (*Mémoires de mathématique et de physique* 1702–1742, vol. 1, 40–44).

The epistemic practice of suspension of judgment formed a coherent methodological whole with patient data collection, hypothetical reasoning, and resistance to framing ambitious systems. Taken together, all of these traits resulted in an attitude that postponed solving a problem or finding an explanation until the future. When natural historians examined snake teeth in 1693, they agreed that they were able to give an 'accurate description'. But when it came to explanations, their methodological scepticism set in. The historians stated that the same thing could serve different purposes, that humans could not penetrate the intentions of God, and that philosophers often erred in their reasoning. But this did not mean that humans should not investigate anything. The members of the research group agreed to conduct further experiments and observations 'that will serve to explain the function of snake teeth in the future' (*Mémoires de l'Académie royale des sciences* 1729–1734, vol. 10, 251). At the beginning of *Mémoires pour servir à l'histoire des plantes,* Dodart wrote that the volume provided an overview of the experiments that academicians conducted and suggested what still needed to be done in the future (Dodart 1679, 6). Much later, in the *HARS* for the academic year 1700, Fontenelle reported on Dodart's research concerning plant breeding. In this case, Dodart also had only presented established facts on the question of plant breeding, but did not yet have a theory (*système*). This required further research and observations (Fontenelle 1702–1742, vol. 2, 65).

Ignorance thus brought not a resigned scepticism that gave up the search for answers but a temporalized scepticism that confidently awaited the answers in the future. When Huygens reported his microscopic investigations to the Académie in 1678, he described the 'infinity of little fish' he saw in a small drop of water. He noted that he observed numerous kinds of 'fish' in various solutions, but he also had no doubt that the cause of this phenomenon would be discovered in time (*Mémoires de l'Académie royale des sciences* 1729–1734, vol. 10, 608).[51] Similarly, Claude Perrault and other authors of the *Mémoires pour servir à l'histoire naturelle des animaux* believed that they articulated only the first tentative speculations to be drawn from the facts they had gathered. When a sufficient number of observations was available, it would be possible to form more accomplished theory from them. Natural historians aimed to proceed from specific observations to general propositions, but the attainment of such a goal would require much work (Perrault 1671, fol. i-r).

In the *HARS* volumes, Fontenelle often stressed that only future research would decide a question to which the members of the Académie did not want to give a definitive answer. For example, in relation to the study of corals, Fontenelle wrote that the interpretation of their growth was 'a question of time and experience' (Fontenelle 1702–1742, vol. 2, 69). Similarly, in anatomy, only the dissection of other bodies would prove the general reasoning of the anatomists of the Académie (Fontenelle 1702–1742, vol. 3, 54). Time was sometimes metaphorically presented as an autonomous epistemic factor that would decide the disputed question: 'We don't know yet. Only time can bring this kind of knowledge.'[52]

Another essential tenet of sage pyrrhonism was the idea that natural philosophy should proceed slowly and gradually. This was not a new idea. Galileo expressed the discrepancy between the divine and human intellect by means of the well-known distinction between the intensive and extensive modes of knowledge. The two intellects differed especially in scope and in speed. While human reason proceeded 'step by step', 'the divine reason, like light, comprehends in an instant.'[53] Maurizio Torrini explained that in Galileo's distinction, 'divine and human knowledge differ only in their time. [...] The temporal dimension belongs only to the sphere of man and his works' (Torrini 2001, 526). God grasped all things immediately and instantaneously, whereas human reason was slow and therefore could only understand nature gradually, slowly, and in parts. In Galileo's view, imperfect human beings

[51] *Mémoires de l'Académie royale des sciences* 1729–1734, vol. 10, 608.

[52] Fontenelle 1702–1742, vol. 7, 37, similarly vol. 9, 25, vol. 10, 36, vol. 11, 49, vol. 12, 57.

[53] Galilei 1890–1909, vol. 7, 128–129, English translation in Galilei 2001, 119; cf. Cassirer 1922, vol. 1, 399–408.

advanced according to the inherent possibilities, and therefore their knowledge advancing in almost imperceptible steps, like climbing a tower.[54]

Bacon also considered gradual and slow development an adequate defence against the excessive haste of reason (Bacon 2004, 160). The apologist of the Royal Society shared this view. According to Glanvill, natural philosophy 'must proceed slowly, by degrees almost insensible' (Glanvill 1668, 91–92). Sprat similarly described the new natural philosophy. Instead of hastily creating fantastic systems and constructing illusory causal relationships, the Royal Society proceeded slowly, following the course of nature and avoiding 'precipitancy' and 'dangerous speed' (Sprat 1734, 154). In Sprat's view, the members of the Royal Society proceeded patiently: 'Their Progress has been equal, and firm, by natural degrees, and thorough small things, as well as great. They go leisurably on; but their slowness is not caus'd by their idleness, but care. (. . .) they should be Judges by what steps, and what pace, they ought to proceed' (Sprat 1734, 318).

Similar expressions can also be found among members of the Académie. According to Cassini, astronomy and geography improved gradually (*se perfectionnoient peu à peu*) (*Mémoires de l'Académie royale des sciences* 1729–1734, vol. 8, 19). Huygens also stated that humans were born in such a way that they had to make their discoveries gradually and over a long period of time in the arts and sciences. No knowledge was suddenly infused into them by God (Huygens 1888–1950, vol. 21, 739). In his dispute with the defenders of the Ancients, Charles Perrault also pointed out that physics advanced degree by degree (*degrez en degrez*) (Perrault 1688–1697, vol. 4, 162).

At the end of the sixth dialogue of the *Entretiens*, Fontenelle wrote: 'When it comes to new discoveries, one should not be in too much of a hurry to reason, even though one is always eager to do so; and true Philosophers are like Elephants, who, while walking, never put the second foot on the ground until the first is well established' (Fontenelle 1990–2001, vol. 2, 128–129). The metaphor of elephants was obviously intended to express the slowness, cumbersomeness, and caution of scientific research. For Fontenelle, natural philosophers were like elephants that slowly, deliberately, and tirelessly followed the path they had set out on. They took their time, gathered facts patiently, and avoided premature conclusions. In the *HARS* volume for the academic year 1709, Fontenelle emphasized that the Académie prided itself on wise slowness (*un sage lenteur*) (Fontenelle 1702–1742, vol. 11, 45).

According to Fontenelle, human reason suffered under a universal urge to rush to conclusions, and natural philosophy was supposed to resist this urge. 'It is necessary to wait until we have a great number of experiences and facts, to draw something general from them; we are commonly in a hurry to establish principles, and the mind runs to the system; but we must not entirely believe this ardour' (Fontenelle and Godin 1733, vol. 1, 62). In Fontenelle's view, human reason had a universal and natural tendency to be subject to various cognitive self-deceptions. The tendency to

[54]Galilei 1890–1909, vol. 7, 479, English translation in Galilei 2001, 527.

hastily create systems was thus a natural and inevitable feature of human reason from which natural philosophers should emancipate themselves.

Natural philosophers, therefore, did not achieve their knowledge immediately and completely as God did. According to Fontenelle, 'knowledge, like everything else, proceeds by stages' (Fontenelle 1702–1742, vol. 4, 45). But philosophers could have control over the coming body of knowledge by setting the pace of research to a value that befitted limited human cognitive abilities. Knowledge of nature thus became the result of autonomous human activity, which depended neither on divine grace nor the anthropocentric setting of nature. Natural philosophers considered the slow pace of their work and deliberate postponing of truth as a mark of its human origin and therefore a natural part of their research. The idea of scientific progress thus did not come to natural philosophy from the outside. It had nothing to do with secularization of eschatology, practices of uneducated artisans, or norms of gentlemanly behaviour. The idea became a means of coping with methodological challenges brought by the new discoveries and reconceptualizations of nature.

4.5 The Infinity of Scientific Progress

All the described features of the temporal regime of knowledge brought the relativization of contemporary knowledge. The empirical basis of knowledge was only provisional, and the same was true for the hypotheses that were based on available data. The set of questions that the hypotheses answered was also provisional. This was an important change because Aristotelian natural philosophy assumed a constant set of questions to be asked, as we know, for example, from the scholastic compendia on cosmology and natural philosophy.[55] At the end of the seventeenth century, natural philosophers expected that they would change their questions about nature according to the shifting of the empirical horizon. It turned out that nobody could establish when this process would end.

Bacon and Descartes were convinced that the progress of natural philosophy would one day end with the restoration of human mastery over nature, definitive morality, or the attainment of wisdom.[56] Their hopes for an predictable early end to scientific progress faded during the seventeenth century. The well-specified goals of knowledge—the wisdom of Adam, *astronomia instaurata*, the sabbath of the sciences, and *sagesse*—became increasingly vague and receded into an indeterminate future. There were several reasons for this loss of definitive goals of scientific inquiry.

The first was the connection of natural philosophy to natural theology, especially in the British context. Some British experimental philosophers placed natural philosophy within an eschatological framework of the history of salvation. Philosophy

[55] See Des Chene 2000; Grant 1994.

[56] Bacon 2004, 22–23, 446–447; Descartes 1996, vol. 9–2, 4.

was meant to be part of the process of restoring Adam's dominion over nature and regaining his faculties and knowledge.[57] The idea of a return to Paradise, however, was gradually abandoned by the members of the Royal Society. They ceased to tie the temporal elements of natural philosophy to eschatology and merged their research with another theological discipline: natural theology. For instance, Hooke wrote that new discoveries made through telescopes and microscopes 'would give us infinite cause, more and more to admire the omnipotence of the Creator.'[58] The expansion of human knowledge was thus represented as an advancement in admiration of the Creator's wisdom and power.

Peter Harrison emphasized that modern historical research often mistakenly considered early modern arguments from design a rational argument in favour of the Christian faith. In fact, natural theology was used to demonstrate the social benefit and religious usefulness of a new experimental philosophy that was a culturally marginalized activity at that time.[59] Closed relationship to natural theology proved that the exploration of nature was not a vain curiosity brazenly penetrating the secrets of God, but a humble affirmation of divine wisdom and power.

I would add that, at least in the British Protestant milieu, natural theology gave legitimacy to inconclusiveness of natural philosophy. Natural theology legitimized the belief in the infinite progress of natural philosophy by declaring nature an expression of the infinite attributes of God. In his *Of the High Veneration Man's Intellect Owes to God* (1684), Boyle wrote of nature: 'that boundless Ocean contains a variety of excellent Objects.' In Boyle's view, there was no doubt that with the further refinement of telescopes our descendants would discover more constellations and stars. 'We need not therefore fear our admiration of God should expire, for want of Objects fit to keep it up.' God is an infinite being and 'progress in the knowledge of the Wonders of the Divine Nature, Attributes and Dispensations', could likewise go in an infinite series of ascending numbers 'farther and farther from the first and least term of the progression [...] without ever reaching or [...] approaching to an Infinite number' (Boyle 1775, vol. 5, 153). The progress of natural philosophy had to be infinite because the divine attributes were also infinite.

In the French context, natural philosophy was much less associated with natural theology,[60] and arguments from design only seldom occurred in the texts of the Académie.[61] What linked the French natural philosophers with their fellows of the Royal Society more closely was the idea that the infinite progress of human knowledge was a consequence of the inexhaustibility of the objects of knowledge.

---

[57] Hamou 2003; Hamou 1999a, vol. 2, 59–70, Harrison 2007.

[58] Hooke 1665, 242; Glanvill 1668, 7; Sprat 1734, 383–388, Charleton 1654, 5; Derham 1738, 29.

[59] Harrison 2008, 264–265. On physico-theology and its significance, see further Harrison 2005, Groh and Groh 1991, 105–120; Armogathe 2007, 251–263.

[60] Gaukroger 2006, 473; Gaukroger 2002, 61–63.

[61] See e.g. *Mémoires de l'Académie royale des sciences* 1729–1734, vol. 10, 703; Fontenelle and Godin 1733, vol. 1, 279.

One of the earliest expressions of the idea of the infinite progress of knowledge was Pascal's preface to his intended treatise on the vacuum, *Préface sur le Traité du vide*, written before Pascal's conversion in November 1654 (the text was published only in 1779). In this short text, Pascal spoke of sciences based on memory, which included things and knowledge already discovered. These sciences, history and theology, were based on authority. Pascal further distinguished sciences based on reason, focusing on the search for hidden truths (*vérités cachées*). Pascal thus separated two dimensions of human knowledge. Memory and its sciences included what was already known. Reason, which dealt with sensible or rational objects, represented the potentiality of knowledge. According to Pascal, the human spirit (*esprit*) could expand without limit: 'Its inexhaustible fertility produces continuously, and its inventions may be multiplied altogether without limit and without interruption.'[62]

In the later *Pensées*, infinity for Pascal was a wistful symptom of the disproportion of humans to the world. In this earlier text, Pascal casually accepted the infinity of knowledge, which he understood as a correlate of its object (Blumenberg 1996a, 743). Hidden truths acted not as a humiliation of humans, but as a challenge that could be mastered with collective cooperation, discovery, and sufficient time. In *Préface*, Pascal was primarily speaking against the authority of the Ancients in natural philosophy, which is why he emphasized the constant presence of hidden truths and the infinite fertility of the spirit. Knowledge could not be fixed within the limits set by the Ancients because the human spirit struggled with infinite nature and its inexhaustible variety. The idea of infinite progress thus appeared in Pascal not as a programme in its own right, but rather as part of an argument against the authority of Greek philosophy. But his text demonstrated that progress could have been considered infinite when it was conceived as a representation of its—infinite—object.

It is not surprising that the transfer of the infinity of nature to natural philosophy can be found in the later quarrel between the Ancients and the Moderns. Despite the limited influence of Pascal's text, which was available only as a manuscript, other authors also used the idea of the inexhaustibility of nature to show that the idea of the insurmountability of ancient models was unjustifiable, at least in the field of natural philosophy. In his *Parallèle des Anciens et des Modernes en ce qui regarde les Arts et la Science* (1693–1697), Charles Perrault stated that 'with the aid of the telescope and the microscope it has been possible to discover an immense space in bodies, both the largest and the smallest, which gives an almost infinite scope to the science that deals with them' (Perrault 1688–1697, vol. 1, fol. **4v).

In his *Préface sur l'utilité des mathématiques et de la physique,* Fontenelle also deduced the infinity of progress from the inexhaustibility of nature and the fertility of the human spirit (*l'esprit*). The various views of the human spirit were almost infinite. Nature was truly infinite (*véritablement*). That was also why new discoveries could be expected every day in mathematics and natural philosophy (*physique*) (Fontenelle 1994, 46). The two disciplines, however, differed as to the possibility of

[62] Pascal 1998, 453; English translation in Pascal 1910, 456.

their completion. Geometry perfected itself every day and eventually would manage to reach perfection. In physics, according to Fontenelle, the situation was different. Physics dealt with a subject whose variety and richness have no limits, and therefore this discipline 'will always find Observations to make, and Occasions of enriching itself, and will keep the Advantage of being a study that can never be completed'.[63]

In Fontenelle's view, the finalization of progress was only possible in the realm of non-empirical human constructs such as mathematics. However, as empirical nature was infinite and inexhaustible, physics would advance forever. The fact that Fontenelle understood this characteristic positively should not be underestimated: it was an 'advantage' that research 'can never be completed'. For him, the idea of reaching the final goal in the empirical disciplines was not only unrealistic but also inappropriate because it implied stagnation in dogmatism and the illusory and deceptive impression of perfect knowledge of infinite nature. In physics, nothing was permanently settled. There always existed room for improvement: even when we think we know nature well, it still eludes us (Fontenelle 1702–1742, vol. 10, 11, cf. vol. 11, 90). When Leibniz introduced binary arithmetic, Fontenelle pointed out that even in mathematics the possibility of further improvement could not be ruled out. What appeared to humans as natural and insurmountable (the decimal system) could be only an illusion of reason, which considered arbitrary ideas necessary.[64]

Next to the transfer of infinity from nature to natural philosophy, another important change led to postponing the final end to the indefinite future: the experience of progress made thus far.

In a short treatise that he read at a meeting of the Académie as early as 1668, Huygens briefly recounted the history of the astronomical tables. The peak of ancient astronomy was Ptolemy's tables, and then the medieval 'dark ages' came. The Renaissance saw a revival of astronomy, but no tables predicted phenomena accurately enough. Huygens admitted that even after centuries of research, astronomers had not been able to produce reliable tables. The effort to calculate tables thus could seem futile. Huygens added, however, that great hope resulted from the achievements of his age, which was in a much better position than preceding periods. There were more observations, better geometry, telescopes, and pendulum clocks. As long as there were instruments and a place for observation (Huygens was referring to the Observatoire), it was reasonable to expect success in this enterprise (Huygens 1888–1950, vol. 19, 262). Huygens's words articulate an epistemological optimism derived from experience with the development of recent natural philosophy. The progress made justified the belief that further progress would follow. Progress began to produce itself.

Cassini held a similar view. In a 1684 report on the discovery of two new satellites of Saturn, Cassini wrote that the continuous progress made in astronomy offered hope that more discoveries would come with time (*Mémoires de l'Académie*

[63] Fontenelle 1994, 316; English translation in Fontenelle 1727, xiv; cf. Fontenelle and Godin 1733, vol. 1, 268; Fontenelle 1990–2001, vol. 2, 419.

[64] Fontenelle 1702–1742, vol. 5, 59, cf. vol. 8, 43.

royale des sciences, vol. 10, 697). In the *HARS* for the academic year 1699, Fontenelle expressed the same conviction. Comparisons with ancient natural history demonstrated the superiority of modern natural history. Even if not everything was yet known, 'the history of progress we have made so far gives hope and courage' (Fontenelle 1702–1742, vol. 1, 52). The sheer intensity of research over the last century offered legitimate hope for further progress. In Fontenelle's view, the more philosophers observed, the more they discovered (Fontenelle 1702–1742, vol. 7, 121).

Thus, the idea of unfinished scientific progress was not promulgated *a priori*. It was the result of a reflection on the development of knowledge in recent decades. As Mazauric observed, the experience with the rapid development of various currents of research at the Académie convinced Fontenelle he could justifiably expect further progress (Mazauric 2007, 35). Exploring nature is not simply a matter of cumulatively adding new knowledge, because new findings require more refined theories. According to Fontenelle, the new and numerous observations in astronomy raised new difficulties that the astronomers of the Observatoire had to solve. A more perfect and accurate understanding of celestial bodies and other objects required more sophisticated hypotheses and solutions than before (Fontenelle 1702–1742, vol. 5, 76, 99, cf. vol. 6, 67–68.). In another volume of *HARS,* Fontenelle used an impressive metaphor: 'Any subject accurately considered becomes infinite and attention is a kind of microscope, which always magnifies and multiplies it, proportionally as it becomes more perfect' (Fontenelle 1702–1742, vol. 8, 15–16; cf. vol. 7, 64). Knowledge of nature found itself in an endless loop of improvement, as better methods led to more discoveries, which resulted in new theories and even better methods leading to further surprising discoveries. Thus, the progress of the sciences was already self-sustaining and did not have to be justified by any ultimate goal.

The last reason for introducing the idea of infinite progress was the idea of the temporariness of all knowledge. The temporal regime of knowledge always linked knowledge to a certain moment in a time continuum that has neither a fixed beginning nor an end. Every moment in the development of knowledge is only a transitional point, preceded by earlier points and followed by later ones. Each further piece of knowledge thus becomes a new start of the process without a definite end.

This awareness of the indefinite end of progress appeared often in the works of academicians involved in the empirical research. In 1673, Cassini, while excitedly announcing the discovery of new planets, also added, that 'various movements of planets will not fail to give the centuries to come ample material for exercise and reasoning' (1673, unpaginated dedication 'Au Roy')." In *Mémoires pour servir a l'histoire des plantes,* Dodart declared his familiarity with the long tradition of natural history and noted that the research of his group represented neither a beginning nor an end. He hoped that curious and gifted people would come who would follow their work in the future (Dodart 1679, 280, 305). Claude Perrault similarly wrote that even after twenty centuries of research conducted by *excellens esprits*, philosophers were still at the beginning, considering the fact that a number of new observations came in every day (Perrault 1680, vol. 3, 9). In *Traité de la lumière*, Huygens also admitted that his hypothesis of light was only a beginning:

'I hope also that there will be some who by following these beginnings will penetrate much further into this question than I have been able to do since the subject must be far from being exhausted'.[65]

The clearly defined goals of progress began to disappear at the horizon of distant future. Instead of the hope of gaining definite knowledge, natural philosophy provided only probable hypotheses with temporary validity. The temporal regime of natural philosophy turned exploration of nature into a continuous improvement without an identifiable goal. The concept of the beginning and the end of the research was thus entirely relativized. Increasingly, the goal of scientific progress became the progress itself. As Blumenberg pointed out, the idea of progress, originally intended to realize an ahistorical ideal, became an end in itself and then changed in an infinite process (Blumenberg 1986, 147). The idea of reaching the definitive knowledge of nature was replaced by endless perfecting.

The academicians used the concept of perfection and perfecting often but they did not explain it. Therefore it is not clear whether they were thinking of endless progress or the attainment of some form of perfection in the future. In his tract from 1693, Cassini described the history of astronomy and geography as a slow, gradual, and continuous progress that consisted of gradual perfection (*Mémoires de l'Académie royale des sciences* 1729–1734, vol. 8, 19). At the end of his text, Cassini promised that astronomy and geography would soon reach the highest perfection (*Mémoires de l'Académie des sciences* 1729–1734, vol. 8, 52Ibid., 52). This remark suggests a belief in a conclusion of the progress of astronomy, but Cassini was less optimistic in his other texts.

In 1690, Cassini wrote that the experience of several centuries had shown that astronomical knowledge improved when astronomers compared large numbers of accurate observations amongst themselves. Approaching the truth, however, had its limits, which astronomers would probably never cross, 'yet it is useful to approach them as near as possible' (*Mémoires de l'Académie royale des sciences* 1729–1734, vol. 10, 315). Like an asymptote, astronomy incessantly approaches ultimate perfection without ever being able to touch it.

Fontenelle believed in the inherent finality of knowledge, which would bring the discovery of general truths and laws. In *Préface sur l'utilité des mathématiques et de la physique*, Fontenelle explained that the research currently going on at the Académie was fragmentary because it was in an early developmental stage after dark centuries of ignorance. But the partial fragments would naturally come together into an ordered whole in the future: 'several distinct Truths, as soon as they are in a sufficient Number, do so lively offer to the Mind their Relations and mutual Dependencies, that it should seem, that after having been separated from each other, by a kind of Violence, they naturally endeavour to be re-united.'[66] Fontenelle seemed to be relying on some kind of teleology. According to this teleological

[65] Huygens 1888–1950, vol. 19, 455; English translation in Huygens 1912, vii.

[66] Fontenelle 1994, 49; English translation in Fontenelle 1727, xvii; cf. Fontenelle 1990–2001, vol. 7, 360, 371.

assumption, knowledge had a natural capacity for self-organization and systematization, as individual truths had an affinity for each other.

Fontenelle seems to have expected a completion of the sciences in the form of a closed system of truths in the distant future. But the text itself does not indicate whether the definitive system is a regulatory idea guiding research or an achievable and realistic goal. In his *Éléments de la géometrie de l'infini* (1727), Fontenelle expressed a similar metaphysical idea through a modified Cartesian metaphor of the tree of knowledge. According to Fontenelle, soon after its birth, every science offered only dispersed truths (*verités dispersées*), which, however, over time began to be linked, though always with difficulty. When enough truths were accumulated, their correspondence appeared and general principles began to emerge. A few small branches would then join together to form larger branches that shaped the trunk of the tree (Fontenelle 1990–2001, vol. 7, 370). Thus scientific knowledge progressed from fragmentary and isolated factual findings to the discovery of general principles. Descartes thought he could derive these principles from his metaphysics at the beginning of his project; according to Fontenelle, principles were the result of a long induction that takes place in the history of science. Whether this induction would ever conclude was not clear from Fontenelle's text.

Bibliography

Anstey, Peter R. 2019. Robert Boyle and the intelligibility of the corpuscular philosophy. In *Experiment, speculation and religion in early modern philosophy*, ed. Alberto Vanzo and Peter R. Anstey, 36–57. New York: Routledge.

Ariew, Roger. 2011. The new matter theory and its epistemology: Descartes (and late scholastics) on hypotheses and moral certainty. In *Vanishing matter and the laws of motion. Descartes and beyond*, ed. Daniel Garber, Dana Jalobeanu, and Peter R.R. Anstey, 31–46. London: Routledge.

Armogathe, Jean-Robert. 2007. *La nature du monde. Science nouvelle et exégese au XVIIe siécle*. Paris: Presses Universitaires de France.

Auzout, Adrien. 1664. *L'Éphéméride du comete*. Paris: Cusson.

———. 1665. *Lettre a M. l'abbé Charles*. Paris: Cusson.

Bacon, Francis. 2004. *The Instauratio magna part II: Novum organum and associated texts*, ed. Graham Rees and Maria Wakely. Oxford: Oxford University Press.

Bloch, Olivier R. 1971. *La philosophie de Gassendi: Nominalisme, matérialisme et métaphysique*. La Haye: Nijhoff.

Blumenberg, Hans. 1985. *The legitimacy of the modern age*, translated by Robert M. Wallace. Cambridge: MIT Press.

———. 1986. *Lebenszeit und Weltzeit*. Frankfurt am Main: Suhrkamp.

———. 1996a. *Die Genesis der kopernikanischen Welt*. Frankfurt am Main: Suhrkamp.

———. 1996b. *Die Legitimität der Neuzeit. Erneuerte Ausgabe*. Frankfurt am Main: Suhrkamp.

Borel, Pierre. 1655. *De vero telescopii inventore*. Hagae: Vlacq.

———. 1656. *Observationum microscopicarum centuria*. Hagae: Vlacq.

Boyle, Robert. 1775. *The works*, ed. Thomas Birch. 6 vols. London: Rivington.

Cassini, Jean-Dominique. 1673. *Découverte de deux nouvelles planetes autour de Saturne*. Paris: Mabre–Cramoisy.

———. 1681. *Observations sur la comete qui a paru au mois de decembre 1680*. Paris: Michallet.

———. 1686. An extract of the journal des Scavans of April 22 ft. N. 1686. Giving account of two new satellites of Saturn discovered lately by Mr. Cassini. *Philosophical Transactions* 16: 79–85.
———. 1810. *Mémoires pour servir à l'histoire des science et à celle de l'Observatoire royal de Paris: suivis de la vie de J.–D. Cassini, écrite par lui–même, et des éloges de plusieurs académiciens morts pendant la révolution*. Paris: Bleuet.
Cassirer, Ernst. 1922. *Das Erkenntnisproblem in der Philosophie und Wissenschaft der neueren Zeit*. 3 vols. Berlin: Olms.
Charleton, Walter. 1654. *Physiologia Epicuro–Gassendo–Charltoniana*. London: Newcomb.
Clarke, Desmond. 1982. *Descartes' philosophy of science*. Manchester: Pennsylvania State University Press.
———. 1989. *Occult powers and hypotheses. Cartesian natural philosophy under Louis XIV*. Oxford: Oxford University Press.
———. 2018. *French philosophy, 1572–1675*. Oxford: Oxford University Press.
Colbert, Jean-Baptiste. 1888. Autre note à Colbert sur l'établissement de l'Académie des beaux–arts et de l'Académie des sciences. In *Lettres, instructions et mémoires de Colbert*, ed. Pierre Clément, vol. 5, 512–515. Paris: Imprimerie impériale.
Comenius, Johannes A. 1964. *De rerum humanarum emendatione consultatio catholica*. 2 vols. Academia: Pragae.
Cotes, Roger. 2004. Editor's preface to the 2nd edition of *Philosophia naturalis principia mathematica* (1713). In Isaac Newton, *Philosophical writings*, ed. Andrew Janiak, 42–59. Cambridge: Cambridge University Press.
Daston, Lorraine. 1998. Probability and evidence. In *Cambridge history of seventeenth–century philosophy*, ed. Daniel Garber and Michael Ayers, 1108–1143. Cambridge: Cambridge University Press.
Dear, Peter. 1985. Totius in verba: Rhetoric and authority in the early Royal Society. *Isis* 76: 146–157.
———. 1995. *Discipline and experience: The mathematical way in the scientific revolution*. Chicago: University of Chicago Press.
———. 2001. *Revolutionizing the sciences. European knowledge and its ambitions, 1500–1700*. London: Bloomsbury Publishing.
Derham, William. 1738. *Astro–theology: Or, a demonstration of the being and attributes of God, from a survey of the heavens (1714)*. London: Innys – Manby.
Des Chene, Dennis. 2000. *Physiologia: Natural philosophy in late Aristotelianism*. Ithaca: Cornell University Press.
Descartes, René. 1996. *Œuvres de Descartes*, ed. Charles Adam and Paul Tannery, 11 vols. Paris: Vrin.
Detel, Wolfgang. 1978. *Scientia rerum natura occultarum. Methodologische Studien zur Physik Pierre Gassendis*. Berlin/New York: De Gruyter.
Dew, Nicolas. 2015. Un Colbertisme scientifique? In *Histoire des sciences et des savoirs. 1. De la Renaissance aux Lumières*, ed. Stéphane Van Damme, 431–445. Paris: Seuil.
Dobre, Mihnea. 2013. Rohault's Cartesian physics. In *Cartesian empiricisms*, ed. Mihnea Dobre and Tammy Nyden, 203–226. Dordrecht: Springer.
———. 2019. Jacques Rohault and Cartesian experimentalism. In *The Oxford handbook of Descartes Cartesianism*, ed. Steven Nadler, Tad M. Schmaltz, and Delphine Antoine-Mahut, 388–401. Oxford: Oxford University Press.
Dodart, Denis. 1679. *Mémoires pour servir a l'histoire des plantes*. 2nd ed. Paris: L'Imprimerie royale.
Donato, Maria P. 2015. Faire corps: Les Académies dans l'ancien régime des sciences (XVIIe–XVIIIe siècle). In *Histoire des sciences et des savoirs. 1. De la renaissance aux Lumières*, ed. Stéphane Van Damme, 87–109. Paris: Seuil.
Du Hamel, Jean-Baptiste. 1701. *Regiae scientiarum academiae historia*. Parisiis: Delespine.
Fontenelle, Bernard de. 1702–1742. *Histoire de l'Académie royale des sciences*. 40 vols. Paris: Boudout.

———. 1727. *Lives of the French, Italian and German philosophers, late members of the Royal academy of sciences in Paris*. London: Innys.
———. 1990. *Conversations on the plurality of worlds*, translated by H. A. Hargreaves. Berkeley: University of California Press.
———. 1990–2001. *Œuvres complètes*, ed. Alain Niderst. Paris: Fayard.
———. 1994. Préface sur l'utilité des mathématiques et de la physique. In Bernard de Fontenelle, *Œuvres complètes*, ed. Alain Niderst, vol. 6. Paris: Fayard.
Fontenelle, Bernard, de, and Louis Godin. 1733. *Histoire de l'Académie Royale des Sciences, depuis son Etablissement en 1666 jusqu'à 1699*. 2 vols. Paris: Martin, Coignard & Guerin.
Gal, Ofer, and Raz Chen-Morris. 2010a. Baroque optics and the disappearance of the observer: From Kepler's *Optics* to Descartes' doubt. *Journal of the History of Ideas* 71: 191–217.
———. 2010b. Empiricism without senses: How the instrument replaced the eye. In *The body as object and instrument of knowledge: Embodied empiricism in early modern science*, ed. Charles T. Wolfe and Ofer Gal, 121–147. Dordrecht: Springer.
Galilei, Galileo. 1890–1909. *Edizione nazionale delle opere di Galileo Galilei*, ed. Antonio Favaro, 20 vols. Firenze: Barbèra Editore.
———. 1989a. *Sidereus nuncius, or the sideral messenger*, translated by Albert van Helden. Chicago/London: The University of Chicago Press.
———. 1989b. *The Galileo affair*, translated by Maurice A. Finocchiaro. Berkeley/Los Angeles/London: University of California Press.
———. 2001. *Dialogue concerning the two chief world systems: Ptolemaic and Copernican*, translated by Stillman Drake. New York: The Modern Library.
Gassendi, Pierre. 1658. *Opera omnia*. 6 vols. Lugduni: Anisson.
Gaukroger, Stephen. 2002. *Descartes' system of natural philosophy*. Cambridge: Cambridge University Press.
———. 2006. *The emergence of a scientific culture: Science and the shaping of modernity, 1210–1685*. Oxford: Oxford University Press.
———. 2010a. The unity of natural philosophy and the end of scientia. In *Scientia in early modern philosophy. Seventeenth-century thinkers on demonstrative knowledge from first principles*, ed. Tom Sorell, G.A.J. Rogers, and Jill Kraye, 19–33. Dordrecht: Springer.
———. 2010b. *The collapse of mechanism and the rise of sensibility. Science and shaping of modernity, 1680–1760*. Oxford: Oxford University Press.
Glanvill, Joseph. 1668. *Plus ultra, or the progress and advancement of knowledge since the days of Aristotle*. London: Collins.
———. 1885. *Scepsis scientifica, or vanity of dogmatizing*, ed. James Owen. London: Kegan Paul, Trench & Company.
Grant, Edward. 1994. *Planets, stars and orbs: The medieval cosmos, 1200–1687*. Cambridge: Cambridge University Press.
Groh, Ruth, and Dieter Groh. 1991. *Weltbild und Naturaneignung. Zur Kulturgeschichte der Natur*. Frankfurt am Main: Suhrkamp.
Guericke, Otto von. 1672. *Experimenta Nova Magdeburgica de Vacuo Spatio*. Amstelodami: Waesberg.
Hacking, Ian. 2006. *The emergence of probability. A philosophical study of early ideas about probability, induction and statistical inference*. Cambridge: Cambridge University Press.
Hahn, Roger. 1971. *The anatomy of a scientific institution. The Paris Academy of Sciences, 1666–1803*. Berkeley/Los Angeles: University of California Press.
Hamou, Philippe. 1999a. *La mutation du visible. Essai sur la portée épistemologique des instruments d'optique au XVIIe siècle*. 2 vols. Paris: Septentrion.
———. 1999b. L'expérience du 'non–pareil' et l'interpretation de la nature selon Robert Hooke. In *Ressemblances et dissemblances dans l'empirisme britannique*, ed. Geneviève Brykman, 11–29. Paris: Centre de Recherche d'histoire de la philosophie moderne et contemporaine.
———. 2003. Ophtalmie, vision adamique et restauration du savoir dans le pensée anglais du XVIIe siècle. *Le temps philosophique* 9: 81–116.

Harrison, Peter. 2005. Physico–theology and the mixed sciences. In *The science of nature in the seventeenth century*, ed. Peter Anstey and John A. Schuster, 165–183. Dordrecht: Springer.

———. 2007. *The fall of man and the foundations of science*. Cambridge: Cambridge University Press.

———. 2008. Religion, the Royal Society, and the rise of science. *Theology and Science* 6: 255–271.

Hevelius, Johannes. 1673. *Machina coelestis*. 2 vols. Gedani: Reiniger.

Hooke, Robert. 1665. *Micrographia, or some physiological descriptions of minute bodies made by magnifying glasses*. London: Martyn and Allestry.

———. 1705. *Posthumous works*. London: Waller.

———. 1726. Discours concerning telescopes and microscopes. In Robert Hooke, *Philosophical experiments and observations*, 257–269. London: Innys.

Huygens, Christiaan. 1888–1950. *Œuvres complètes*. Vol. 22 vols. La Haye: Nijhoff.

———. 1912. *Treatise on light*, translated by S. P. Thompson. London: MacMillan London.

Jelaco, Ron. 2016. Adrien Auzout and the origins of the Paris observatory. In *Chora 7: Intervals in the philosophy of architecture*, ed. Alberto Pérez-Gómez and Stephen Parcell, 117–140. Montreal: McGill–Queen's University Press.

Joy, Lynn S. 2006. Scientific explanation from formal causes to laws of nature. In *The Cambridge history of science. Vol. 3: Early modern science*, ed. Lorraine Daston and Katherine Park, 70–105. Cambridge: Cambridge University Press.

Lambert, Ladina B. 2002. *Imagining the unimaginable. The poetics of early modern astronomy*. Amsterdam/New York: Rodopi.

Laudan, Laurens. 1966. The clock metaphor and probabilism: The impact of Descartes of English methodological thought 1650–1665. *Annals of Science* 22: 73–104.

Leeuwenhoek, Anthoni van. 1939–1999. *Alle de brieven*. 15 vols. Amsterdam: Swets and Zeitlinger.

Locke, John. 1975. *Essay concerning human understanding*, ed. Peter H. Nidditch. Oxford: Clarendon Press.

Marchal, Roger. 1997. *Fontenelle à l'aube des lumières*. Paris: Honoré Champion.

Mariotte, Edme. 1678. *Essay de Logique*. Paris: Michallet.

Marsak, Leonard Mendes. 1959. Bernard de Fontenelle: The idea of science in the French enlightenment. *Transactions of the American Philosophical Society* 49: 1–67.

Mazauric, Simone. 2007. *Fontenelle et l'invention de l'histoire des sciences à l'aube des Lumières*. Paris: Fayard.

———. 2009. *Histoire des sciences à l'époque moderne*. Paris: Armand Colin.

McClaughlin, Trevor. 1977. Le concept de science chez Jacques Rohault. *Revue d'histoire des sciences* 30: 225–240.

McKeon, Robert M. 1965. Une lettre de Melchisédech Thévenot sur les débuts de l'Académie royale des sciences. *Revue d'histoire des sciences* 18: 1–6.

McMullin, Ernan. 2008. Explanation as confirmation in Descartes's natural philosophy. In *A Blackwell companion to Descartes*, ed. J. Broughton and J. Carriero, 84–102. Oxford: Oxford University Press.

———. 2009. Hypothesis in early modern science. In *The significance of the hypothetical in the natural sciences*, ed. Michael Heidelberger and Gregor Schiemann, 7–38. Dordrecht: Springer.

Mémoires de l'Académie royale des sciences, depuis 1666 jusqu'à 1699. 1729–1734. 11 vols. Paris: Compagnie des Libraires.

Mémoires de mathématique et de physique, tirez des registres de l'Académie royale des sciences. 1702–1742. 40 vols. Paris: Boudot (the same volumes as Fontenelle 1702–1742 but with separate pagination).

Nadler, Steven. 2000. Doctrines of explanation in late scholasticism and in the mechanical philosophy. In *The Cambridge history of seventeenth–century philosophy*, ed. Daniel Garber and Michael Ayers, vol. 1, 513–552. Cambridge: Cambridge University Press.

Newton, Isaac. 2004. *Philosophical writings*, ed. Andrew Janiak. Cambridge: Cambridge University Press.
Osler, Margaret. 1970. John Locke and the changing ideal of scientific knowledge. *Journal of the History of Ideas* 31: 3–16.
Pascal, Blaise. 1910. Preface to the treatise on vacuum, translated by O. W. Wight. In Blaise Pascal, *Thoughts. Letters. Minor works*, ed. Blaise Pascal, 440–450. New York: Collier.
———. 1998. Préface sur le Traité du vide. In Blaise Pascal, *Œuvres complètes*, ed. Michel Le Guern, vol. 1. Paris: Gallimard.
Pasnau, Robert. 2017. *After certainty. A history of our epistemic ideals and illusions*. Oxford: Oxford University Press.
Perrault, Claude. 1671. *Mémoires pour servir à l'histoire naturelle des animaux*. Paris: L'imprimerie royale.
———. 1680. *Essais de physique*. 4 vols. Paris: Coignard.
Perrault, Charles. 1688–1697. *Parallèle des anciens et des modernes*. 4 vols. Paris: Coignard.
Petit, Pierre. 1665. *Dissertation sur la nature des comètes*. Paris: Billaine.
Picolet, Guy. 2021. L'Académie avant l'Académie: les premières activités de la compagnie en formation (mai–décembre 1666). *Dix–septième siècle* 292: 71–105.
Popkin, Richard. 2003. *The history of scepticism. From Savonarola to Bayle*. Oxford: Oxford University Press.
Régis, Pierre-Sylvain. 1690. *Système de philosophie, contenant la logique, la métaphysique, la physique et la morale*. Paris: Anisson.
Rheita, Schyrlaeus. 1645. *Oculus Enoch et Eliae sive Radius sidereomysticus*. 2 vols. Antverpiae: Verdussius.
Rogers, G.A.J. 2010. John Locke and the limits of *Scientia*. In *Scientia in early modern philosophy. Seventeenth–Century thinkers on demonstrative knowledge from first principles*, ed. Tom Sorell, G.A.J. Rogers, and Jill Kraye, 129–136. Dordrecht: Springer.
Rohault, Jacques. 1671. *Traité de physique*. Paris: Thierry.
Roux, Sophie. 1998. Le scepticisme et les hypothèses de la physique. *Revue de Synthèse* 119: 211–255.
———. 2011. *L'Essai de logique de Mariotte. Archéologie des idées d'un savant ordinaire*. Paris: Garnier.
———. 2013a. An empire divided: French natural philosophy (1670–1690). In *The mechanization of natural philosophy*, ed. Daniel Garber and Sophie Roux, 55–96. Dordrecht: Springer.
———. 2013b. Was there a Cartesian experimentalism in 1660s France? In *Cartesian empiricisms*, ed. Mihnea Dobre and Tammy Nyden, 203–226. Dordrecht: Springer.
———. 2017. The two comets of 1664–1665: A dispersive prism for French natural philosophical principles. In *The idea of principles in early modern thought*, ed. Peter R. Anstey, 98–147. London: Routledge.
Sacco, Francesco G. 2020. *Real, mechanical, experimental. Robert Hooke's natural philosophy*. Cham: Springer.
Sakellariadis, Spyros. 1982. Descartes' use of empirical data to test hypotheses. *Isis* 73: 68–76.
Seguin, Maria Susana. 2006. Rhétorique scientifique et épistémologie dans les *Mémoires de l'Académie des sciences*. *Revue Fontenelle* 4: 147–159.
———. 2012. Fontenelle et l'Histoire de l'académie royale des sciences. *Dix–huitième siècle* 44: 365–379.
———. 2023. The Royal academy of sciences of Paris and the birth of historical epistemology. In *Ordering knowledge: Disciplinarity and the shaping of European modernity*, ed. Jean-Jacques Chardin, 187–208. Strasbourg: Presses universitaires de Strasbourg.
Serjeantson, Richard. 2006. Proof and persuasion. In *The Cambridge history of science. Vol. 3: Early modern science*, ed. Lorraine Daston and Katherine Park, 132–175. Cambridge: Cambridge University Press.
Shank, John B. 2018. *Before Voltaire: The French origins of 'Newtonian' mechanics, 1680–1715*. Chicago/London: The University of Chicago Press.

Sorbrière, Samuel. 1917. Discours de Sorbrière à l'ouverture de l'Académie de Montmor, ed. Georges Bigourdan. *Comptes Rendus de l'Académie des Sciences* 164: 159–162, 216–220.

Specht, Robert. 2009. Experiences and hypotheses: Opinions within Locke's realm. In *The significance of the hypothetical in the natural sciences*, ed. Michael Heidelberger and Gregor Schiemann, 38–58. Berlin/New York: De Gruyter.

Sprat, Thomas. 1734. *The history of the Royal society of London*. London: Knapton.

Stroup, Alice. 1990. *A company of scientists: Botany, patronage, and community at the seventeenth-century Parisian Royal académie of sciences*. Berkeley/Los Angeles: University of California Press.

Sturdy, David. 1995. *Science and social status: The members of the Académie des sciences (1666–1750)*. Woodbridge: The Boydell Press.

Tits-Dieuaide, Marie-Jeanne. 1998. Les savants, la société et l'État: à propos du renouvellement de l'Académie royale des sciences (1699). *Journal des savants* 1: 79–114.

Torrini, Maurizio. 2001. The time of men and the time of objects: Galileo and his contemporaries. In *The medieval concept of time*, ed. Pasquale Porro, 519–529. Leiden: Brill.

Tournefort, Joseph P. 1698. *Histoire de plantes quie naissent aux environs de Paris*. Paris: L'Imprimerie Royale.

Van Helden, Albert. 1976. The importance of the transit of mercury of 1631. *Journal for the History of Astronomy* 7: 1–10.

———. 1996. Contrasting careers in astronomy: Huygens and Cassini. *De Zeventiende Eeuw* 12: 96–105.

Varignon, Pierre. 1690. *Nouvelles conjectures sur la pesanteur*. Paris: Jombert.

Waard, Cornelis, de. 1925. Les objections de Pierre Petit contre le Discours et les Essais de Descartes. *Revue de Métaphysique et de Morale* 32: 53–89.

Wade, Ira O. 1977. Fontenelle and the philosophical spirit. In *The structure and form of the French enlightenment, vol. 1: Esprit philosophique*, 26–34. Princeton University Press.

Walsh, Kirsten. 2019. Newton's scaffolding. The instrumental roles of his optical hypotheses. In *Experiment, speculation and religion in early modern philosophy*, ed. Alberto Vanzo and Peter R. Anstey, 125–157. New York: Routledge.

Westfall, Richard. 1956. Unpublished Boyle papers relating to scientific method. *Annals of Science* 12 (63–73): 103–117.

Wilson, Catherine. 2019. What (else) was behind the Newtonian rejection of 'Hypotheses'. In *Experiment, speculation and religion in early modern philosophy*, ed. Alberto Vanzo and Peter R. Anstey, 159–183. New York: Routledge.

Wolf, Charles. 1902. *Histoire de l'Observatoire de Paris de sa fondation à 1793*. Paris: Gauthier–Villars.

Chapter 5
Ars Longa, Vita Brevis: Collectivization of Knowledge

Contents

Abstract Since classical antiquity, gathering natural knowledge was a task meant for an individual. Renaissance natural historian Conrad Gessner and astronomer Tycho Brahe also privileged the individual and private character of natural knowledge. Both scientists considered their research closed work to be passed on to grateful posterity, providing them with literary immortality. During the seventeenth century, it became clear that no individual experience could suffice to produce a reliable system of knowledge. Natural philosophers started to use several tropes and metaphors that were meant to express the insufficiency of ancient natural philosophy and a belief in the necessity of collaboration over generations. At the end of the seventeenth century, institutions such as the Académie des sciences and the Observatoire of Paris supported the development of collective epistemic practices, highlighting the fact that proper research of nature demanded an institutional background. The enduring institution could give the research long-term continuity, outlive the short lifetimes of the individuals, and preserve data, ideas, and theories in its registers. Thus, scientific knowledge was given a new collective bearer: humanity in general, and scientific institutions in particular.

Keywords Conrad Gessner · Tycho Brahe · Jean-Dominique Cassini · Académie des sciences · Early modern academies · Scientific institution · Bernard de Fontenelle

The descriptive idea of progress was based on the comparison of old and new knowledge, which led to the assertion of the superiority of the present. When the temporal regime of knowledge entered natural philosophy, natural philosophers also

D. Špelda, *The Origins of the Idea of Scientific Progress*, International Archives of the History of Ideas Archives internationales d'histoire des idées 250,
https://doi.org/10.1007/978-3-031-60526-0_5

began to be concerned with the problem of who would guarantee the continuity of progress in the future (Neuser 2011, 791b). Realizing that the infinite variety of nature exceeded the cognitive capacity of an individual life, natural philosophers concluded that they would need a much longer lifetime to understand all of nature. Boyle and Ray, for example, agreed that natural philosophy could investigate so many objects that even if 'a man had the age of Methuselah to spend', it would not be enough to know nature (Boyle 1775, vol. 2, 10; cf. Ray 1735, 369). Another way of expressing the disproportion between the brevity of life and the inexhaustibility of nature was the aphorism supposedly formulated by Hippocrates in classical antiquity: *ars longa, vita brevis* (Hippocrates 1844, 458).

Richard Yeo wrote that early modern England offered two main solutions to Hippocrates's complaint: prolonging short human lives or 'imagining individual lives added together over generations to create a *vita longa*' (Yeo 2018, 78–79). The belief in prolonging life assumed that the bearer of knowledge would remain the individual, who would be improved by more developed medicine. The second option presupposed the possibility of unifying individuals into a collective entity passing through the history. Yeo did not deal with a third solution to the disparity between human life and nature's time: creating a collective *vita longa* by means of institutionalization. The subject of knowledge was to become a collective entity, an institution capable of unifying, accumulating, and preserving the experience of individuals.

In this chapter, I consider the second and third options. Using Renaissance astronomy and natural history as examples, I first show that the collectivization of knowledge was far from a natural epistemic stance. Next, I want to present the first attempts to create a collective bearer of knowledge through metaphors. Finally, I will discuss the institutionalization of the sciences in the Académie des sciences. I believe that the creation of research continuity through the institution represented another condition of possibility of the idea of scientific progress.

5.1 The Problem of a Short Lifetime in Astronomy

The first references to the incommensurability of time in human life and nature can be found in classical antiquity, as Blumenberg observed.[1] The Greek astronomers found out that some phenomena in the heavens, such as the precession of the equinoxes, change so slowly that to discern and understand them, empirical data had to be collected over centuries.[2] Even if an individual astronomer summoned all their strengths, the short time interval of their observations would be insufficient to discover or recognize these phenomena. Astronomy, therefore, required the long-term cooperation of astronomers from various times.

[1] Blumenberg 1986, 99–129.

[2] Ptolemy 1998, 329; cf. Blumenberg 1986, 108–110.

As Blumenberg further observed (1986, 112), the brevity of human life also became apparent when confronted with rare phenomena such as comets. Blumenberg referred to the seventh book of Seneca's *Naturales Quaestiones*, where the Roman philosopher discussed comets. While Aristotle judged that comets were meteorological phenomena, Seneca was convinced that comets were celestial bodies that periodically returned to Earth. Humans were ignorant of the laws of their motion since comets were an unusual spectacle and they returned after a long time.

> There will come a day when the passage of time and the efforts of a longer stretch of human history will bring to light things that are now obscure. One lifetime, even if it can be wholly devoted to astronomy, is not sufficient for the investigation of such important matters. [. . .] So it will take a long succession of people to explain these matters. There will come a day when our descendants are astonished that we did not know such obvious facts (Seneca 2014, 130).

Seneca realized that rare phenomena could not be explored in a single lifetime. Their exploration had to be stretched over time by means of a consciously constructed tradition. Because of their infrequency (*propter raritatem eorum*) 'it is essential to have records of past appearances of comets' (Seneca 2014, 116). Understanding the trajectories of comets was thus a task for generations of researchers who could pass on their knowledge over time in a comprehensible and methodically organized manner.

Slow and rare astronomical phenomena provided evidence that, despite the assurances of Greek philosophers, the cosmos was not completely accessible to the human desire for knowledge. Investigations of the universe demanded a long research tradition. This belief was held by astronomers until the Renaissance, but it should not be understood as an idea of progress. In Renaissance times, the long duration of astronomy was seen as a time for the accumulation of errors and inaccuracies that diminished the value of the accumulated observations.

At the time of the Renaissance, there was a widespread belief that astronomy was in crisis. Astronomers were convinced that errors, mistakes, bad constructions, and wrong conclusions had been accumulating in their field since classical antiquity. They wanted to purge astronomy of such errors and return it to the original state that Greek astronomy had embodied. When astronomers made changes to astronomical theories, they believed that they were returning astronomy to its original intact state from the beginning of history. That is why they liked to denote their research as *restitutio, restauratio, emendatio*, or *instauratio*, in an attempt to convey that they were purifying the field and returning to the original source (Rose 1975, 1–3, 90–91; Špelda 2013).

Copernicus regarded even his heliocentric astronomy as a restoration of the original state. He praised Ptolemy, saying that the Greek astronomer 'brought this entire art almost to perfection' by using observations gathered for more than 400 years. In Ptolemy's time, there 'no longer seemed to be any gap which he had not closed'.[3] In Copernicus's view, inaccuracies and shortcomings accumulated after Ptolemy over time.[4]

According to Copernicus, the errors emerged partly due to the short duration of the astronomical tradition on which Ptolemy's *Almagest* was based. Copernicus believed that he could dare to reform astronomy because he had observations that spanned much longer periods of time and thus allowed a better understanding of the nature of particularly slow and rare phenomena.[5] Joachim Rheticus, the only disciple of the Polish astronomer, summed up Copernicus's superiority to Ptolemy simply by saying that Copernicus had observations for 2000 years, while Ptolemy had data for barely a quarter of that period (Rheticus 1982, 52).

The accumulated experience and the time gap between observations led Copernicus to the conviction that he could rid astronomy of errors, eliminate monstrous constructions, and restore the discipline to its original dignity, which consisted of knowing 'the true form of the world and the symmetry of its parts' (Copernicus 1975, 4). Copernicus thus seemed to believe that his work meant the end of the history of astronomy. Two thousand years was long enough to produce a true astronomy that corresponded to the phenomena (Blumenberg 1986, 117, cf. Blumenberg 1996, 645). What he handed down to posterity in his *De Revolutionibus* (1543) was to be more or less finished astronomy, not an incentive to further exploration and improvements. Copernicus's work was intended to be a mature synthesis replacing Ptolemy's *Almagest* and allowing only minor corrections in future. From Rheticus, we know that Copernicus did not anticipate successors to improve and complete his unfinished work. They were merely expected to confirm it (Rheticus 1982, 49).

Tycho Brahe comprehended his astronomical research similarly. Between 1576 and 1597, the Danish astronomer carried out long-term and systematic research at his astronomical observatory Uraniborg on the island of Hven, which produced the largest and most accurate collection of observational data in the history of astronomy to date. Tycho also claimed that he was trying to remove the errors from astronomy. He therefore presented his programme as a restoration (*instauratio*) of astronomy to its former glory and dignity.[6]

Due to the complexity of the observational activities that Tycho wanted to accomplish, he could not carry out the observations alone. We know from the works of modern historians such as Victor Thoren and John R. Christianson that Tycho did not spend much time on observations because he was occupied with writing books and letters or organizing empirical data. The observing instruments were operated by teams of his students and assistants, to whom Tycho provided food, lodging, education, or salary. Tycho viewed these people as part of his broad scholarly household, which he ruled as a patriarch with unquestioned authority

[3]Copernicus 1975, 8, English translation in Copernicus 1978, 8.

[4]Rheticus 1982, 52, English translation in Rheticus 1971, 132.

[5]Copernicus 1975, 8; English translation in Copernicus 1978, 8; cf. Blumenberg 1965, 357–359.

[6]Brahe 1913–1929, vol. 1, 148; cf. vol. 2, 150–151, 301, 304; vol. 5, 5–6.

(Thoren 1990, 116, 201–204; Christianson 2000, 58–76, cf. Gingerich and Voelkel 1998). The Uraniborg observers were therefore not Tycho's equals, but assistants, students, subordinates, and employees. After all, many of the incoming students were at the observatory only temporarily. They saw their internship at Uraniborg as part of their education on their way to studying theology or medicine. In later life, they mostly paid only marginal attention to astronomy (Thoren 1990, 192–207; Christianson 2000, 58, 98, 80–91, 125, 143–145, 150–153).

Tycho had an ownership interest in the results and resources of the Uraniborg research. He regarded the observatory building and its equipment as his private property, in which he had invested heavily (Brahe 1913–1929, vol. 5, 5–6, 14, 30–31, 41–42, 58, 104). Tycho also considered the collected observational data to be his private property and the students had only limited access to it. He 'guarded' his data as a 'most precious and valuable treasure' and hoped, if God's grace permitted, to add more observations to them and perhaps (*forte*) present them to the public.[7] In fact, he was in no hurry to publish his data, which were not made available until the seventeenth century, long after his death (Van Helden 1994).

Tycho described the *instauratio astronomiae* as 'an immense and difficult task' that he grasped as his personal mission. He did not count the students and staff who worked with him on Hven in the completion of the project. Tycho was convinced that he could accomplish the enormous task of reforming astronomy on his own by intensifying his efforts and using all his resources: buildings, instruments, employees, students, and royal support.

At the same time, he presented the instauration of astronomy as the final act by which astronomy would be restored, and thus its development would end. Astronomy could 'finally be restored to wholeness and handed down to posterity more correctly than any time before'.[8] Similarly to other Renaissance authors (Ianziti 2012, 258–259; Zaho 2004, 33), Tycho wanted to pass his work on to posterity as his personal legacy. The result of his mission should be a finished work intended as a gift that would ensure his eternal glory. The Danish astronomer expected future astronomers to look upon him with admiration and gratitude.[9]

Tycho styled his position in the history of astronomy according to this view of his own work. Adam Mosley[10] has drawn attention to an anonymous *Synopsis* that was appended to *Epistolae Astronomicae* (1596). Here the author wrote that Tycho had a room in Uraniborg with eight large portraits. Six of them belonged to great astronomers of the past, such as Ptolemy or Copernicus. The seventh portrait in the row was of Tycho himself. The eighth depicted Tychonides, his son. Under Tycho's portrait, it was written that his astronomical work would be judged by posterity. The inscription below Tychonides's image in turn communicated that the next generation

[7]Brahe 1913–1929, vol. 5, 110; cf. Mosley 2007, 99, 237–238; Christianson 2000, 151–152.

[8]Brahe 1913–1929, vol. 5, 9, English translation in Brahe 1996, 10.

[9]Brahe, letter to Caspar Peucer, 13 September 1588, Brahe 1913–1929, vol. 7, 132–133; cf. vol. 2, 281; vol. 5, 120, 126; cf. Dreyer 1890, 162–164, 182–183; Mosley 2007, 110, 132, 138.

[10]Mosley 2007, 25, 69, 132 commenting on Brahe 1913–1929, vol. 6, 274–275.

would continue to carry Tycho's Atlas-like plight (*Atlantum continuare vices*) (Brahe 1913–1929, vol. 6, 274–275).

This iconography reveals to some extent how Tycho positioned himself in the history of astronomy as a successor of respectable predecessors. At the same time, it also gives a hint of how Tycho envisioned two essential aspects of the future of his field. The first aspect concerned the continuation of research. Tycho seemed to expect succession and continuation of the work in the direct family tradition, not from his employees and disciples. Tycho thought of his observatory as a feudal fiefdom to be passed on to his heirs, rather than as a collective body to compensate for the insufficiency of an individual life. The *vita longa* necessary for the collection of observations was to be provided by direct succession from father to son.

The second aspect concerned the possibilities for further development of astronomy. Tycho seemed to hope that the respect of descendants would also be reflected in continuing his project. The research was to be carried out by the same means and methods that he had developed himself. Tycho did not presuppose a change in astronomical agenda, methods, or goals. Rather, he believed that posterity would fill in the missing data without asking new questions or posing new problems. Thus, Tycho's work undoubtedly contains the descriptive idea of progress, but the belief in the collective extension of astronomical research into the future is missing. Without such a future perspective, the temporal regime of knowledge, and thus the idea of scientific progress, could not be established.

5.2 Information Overload in Natural History

Recent research has shown that Renaissance natural history was primarily oriented towards collecting findings from literary sources and transforming them into a new arrangement.[11] Despite their literary and humanistic orientation, natural historians also sought to grasp and adapt new experiences—especially descriptions of new animals and plants that came from northern and central Europe and the Americas and other continents. Recent historians have metaphorically referred to the new experiences coming to European naturalists of that time as 'information overload'. The flood of reports of previously undescribed plants and animals led Brian W. Ogilvie to the conclusion that 'natural history had changed radically: from a closed world it had become an almost infinite universe' (Ogilvie 2003, 33, cf. Blair 2003).

It is not difficult to find support for this claim. In the Latin preface to his *Histoire des Plantes* (1577), Rembert Dodoens lamented the immeasurable number of plant species scattered over the world (Dodoens 1577, fol. *iiv). In the preface to the third volume of his *Historia Animalium* (1555), Conrad Gessner explained that natural history as a 'knowledge of particulars' struggled to come to terms with the great

[11] Riedl-Dorn 1989; Ashworth 1990; Blair 1992; Grafton 1992; Findlen 2006; Harrison 2011.

number and variety of natural objects.[12] In his *Histoire Entière des Poissons* (1558), Guillaume Rondelet also complained that natural particulars were 'vast, even infinite in number' and that even the lives of several humans were not enough to know them (Rondelet 1558, fol. a4r).

Information overload had similar epistemological effects in natural history as slow and rare phenomena had in astronomy, because it demonstrated the disproportion between the capacity of the individual life and the diversity of nature. Encyclopaedic projects collecting literary erudition seemed feasible at the individual level. However, experience with the radical growth of experience made clear that any individual would find it difficult to achieve the ambition of creating an all-encompassing description of nature, or even just one of its parts, such as fish.

Renaissance natural history developed two strategies to deal with the problem of the immense diversity of nature in the individual lifetime. The first strategy was to neutralize the novelty of the plants and animals recently discovered. The second focused on the intensification of individual effort.

Despite the existence of information overload, some natural historians were convinced that the ancient conceptual paradigm dividing the world of living nature into various genera was sufficient to accommodate the flood of new species (Leinkauf 2017, 1403–1404, 1425). In his 1555 natural history of birds, Pierre Belon rejected the notion that there was an uncountable number of species. In his view, God gave humans reason and wanted it to be within their capacities to count the things that were 'created for their use'. Belon believed that humans could reduce countless singular natural objects to a few hundred species of birds, quadrupeds, reptiles, and plants (Belon 1555, 66). Rondelet insisted on the lasting validity of Aristotelian philosophical principles and categories, into which later generations of Greek naturalists inserted further insights that had not survived in their entirety (Rondelet 1558, fol. a4r).

In his 1558 description of American nature, André Thévet acknowledged that the Ancients did not know America (Thévet 1558, fols. 34v–36v), but at the same time, he repeatedly referred to them 'to minimize the newness of the New World', as Andrea Frisch explained (Frisch 2014, 64). In a recent article, Jana Černá and Lucie Čermáková used many examples to show that Central European and Spanish natural historians understood newly discovered plants as analogous to plants described in ancient natural history (Čermáková and Černá 2018). Natural historians thus diminished the troubling diversity of nature by placing new species within the paradigm of Greco-Roman natural history, which continued to play two roles: It worked both as methodological example and treasury of the known.

The conviction of the exemplarity of Greek knowledge was also based on the fact that Renaissance culture was never sure whether its discoveries and works were not already known in classical antiquity and just forgotten.[13] Believing in the essential

---

[12]Gessner 1551–1555, vol. 3, fol. α3v: *in hac rerum singularium infinita ferme copia ac varietate*; cf. similarly Oviedo 1555, fol. b–iv.

[13]Cardano 1663, vol. 2, 548a; Panciroli 1655, 1; Vergil 2002; cf. Copenhaver 1978.

intelligibility of nature, Renaissance natural historians often assumed that the Ancients had already known almost all of what humans were allowed to know (Müller 1998, 295). The silence of ancient sources concerning new natural objects from northern Europe or the Americas was interpreted as the presence of gaps in ancient natural history. The gaps occurred because the process of *translatio studii* had been unreliable and fragmentary. Therefore, the idea of modern superiority could easily be dismissed as an illusion caused by cultural entropy.

For this reason, German botanists, as Jan-Dirk Müller emphasized, had a strong tendency to interpret new discoveries as a rediscovery of something that had been known in classical antiquity. The widening of human knowledge was presented as a return to the ancient authors: to their knowledge and their experience, which remained the authority for coming empirical findings (Müller 1991, 130–133). The same attitude can be found in Belon who recognized the experience recorded by the Greeks as model according to which he interpreted his own observations.[14] Bodin and Gessner similarly assumed that they did not possess the Greek and Roman knowledge in its completeness. They considered their task to expose, clarify, and explain the knowledge of their ancient predecessors and to fill the lacunae remaining in classical works.[15]

These examples show that for many Renaissance natural historians, most new objects from American nature had their antecedents or analogies in ancient knowledge, since this knowledge was assumed to encompass almost everything that could be known. Natural history was thus able to neutralize the unsettling tide of new experience by inserting it in the traditional structures defining the familiar world. The new experience thus served neither as an impetus for emancipation from the ancient tradition nor as an opening into new ways of research. For that reason, natural historians had no reason to develop the prescriptive idea of progress.

The second strategy for coping with the infinite extent of nature focused on the individual. Gessner, while pointing to the diversity of nature, also harboured the hope that the knowledge of nature could be accomplished in the individual lifetime if enough strength and resources could be mustered. Gessner believed that if he could find suitable patrons, the work would be *multo absolutius* (Gessner 1551–1555, vol. 3, fol. α4r). On the other hand, conscious of the shortness of life, Gessner hastened his work (Müller 1986, 335). As Udo Friedrich observed, Gessner tried to speed up his research through lists, indexes, and notes intended to increase the economy of reading other authors' texts and writing his own (Friedrich 1995, 55). Such aids, however, did not represent an universal programme or methodology, but only the individual facilitation of a difficult task. Gessner did not want to achieve a collective *vita longa*, but to make his *vita brevis* more effective.

Gessner was realistically aware that the completion of a work of natural history could not be achieved without collaboration. In a letter written in 1556, Gessner

[14] Belon 1551, fol. 24v; quoted by Pinon 2003, 55; cf. Pinon 1995, 91.

[15] Bodin 1597, 6; cf. Gessner 1551–1555, vol. 1, fol. β3r; Rondelet 1558, fol. a4r; cf. Ogilvie 2006, 88–103; Graf-Stulhofer 2012.

complained that the vast number of plant species exceeded the capacity of the individual: *unus vir, nullus vir*. Yet Gessner optimistically believed that if his contemporaries cooperated, it would be possible to complete the natural history of plants in one great comprehensive work.[16]

By collaboration, however, Gessner did not mean collective research; he followed the model of antiquity, as Müller demonstrated (Müller 1986, 332–333). That model was Pliny's account of Alexander the Great's commissioning all the artisans and naturalists of his empire to send their observations to Aristotle (Pliny 1967, 35). Like Aristotle, Gessner stated that he tried to gather as much experience as possible in his travels in Europe, drawing on as many authors as possible, interviewing uneducated people, artisans, travellers, etc. (Gessner 1551–1555, vol. 1, fols. α2v, α3v; cf. vol. 3, fol. α4r). However, this collection of experiences gathered around the world had a single mind to evaluate, process, and systemize data into one conceptual structure (Pinon 2003, 64). Gessner did not seem to envision a collective collaboration dividing the research tasks among members of a team. In his view, an individual author stands behind a systematic body of natural knowledge.

Both strategies resulted in a specific view of further development of natural history which was far from the idea of scientific progress. Natural historians such as Gessner and Rondelet were similar to astronomers in their attitude to the future of their work. They also thought they were submitting complete and more or less finished work. At the end of the second preface to the first volume of the *Historia animalium,* Gessner expressed the hope that his work would endure for all posterity, not through the merit of his personal scholarship, which was slight, but through the diligence with which he had carefully combined so much knowledge from countless authors into a single body (*thesaurum*).[17] Similarly to Tycho, Gessner implied that he was handing down a finished work, a compilation of the known—for which he expected gratitude and admiration from posterity.

Although he admitted the inconclusiveness of his text (*cui semper accedere multa possent*) (Gessner 1551–1555, vol. 1, fol. β3r), Gessner seemed to have pictured an extension of his work along the same lines as he did his research—followers would perform a mere *accedere*. In the same text, Gessner wrote that all subsequent descendants would use his writings (*ad omnem posteritatem duraturum hunc laborem nostrum*). Gessner might have hoped that his work would remain a permanent source of knowledge for the entire future because the paradigm of living nature would remain the same without introducing new problems and new questions. As the ambition of works of Gessner and other naturalists was to replace the ancient models, their natural history was to remain in, as a closed thesaurus of knowledge that could be added to but not structurally changed. Gessner's conception of the

---

[16] Gessner 1577, vol. 3, fol. 137v; cf. Pinon 2005, 247–248.

[17] Gessner 1551–1555, vol. 1, fol. β3r: *Spero enim, absit invidia, ad omnem posteritatem duraturum hunc laborum nostrum, non quidem doctrinae nostrae, quae exigua est, sed diligentiae merito, quae tot tantasque lucubrationes ex innumeris authoribus in unum quasi thesaurum accuratissime coniuxit.*

future of natural history, therefore, seems not to include the idea that his own work might become obsolete and be surpassed.

The descriptive idea of progress was easily compatible with this approach. Friedrich noted that Gessner, in the preface to the Latin edition of Hieronymus Bock's *Kräuterbuch* (1552), outlined his progressive view of the history of plant healing: after centuries of barbarism, knowledge of the healing powers of plants had been restored and had reached a level that probably no previous century had achieved.[18]

In Friedrich's interpretation, Gessner was convinced that a new epoch was coming that not only surpassed the previous one but also made itself independent of it by adding its own discoveries. Gessner also hoped this process would continue (*maiorum inventis pauca semper paulatim adijciuntur*) (Friedrich 1995, 55). I would not confuse Gessner's standpoint with the prescriptive idea of scientific progress. He may have believed that his age represented a superior epoch, but given the quality of knowledge, it was merely an accretion. The present generation was adding a little knowledge (*pauca*) to what the Ancients had discovered, whose paradigm was the starting point and oriented the present experience as well.

Dodoens also articulated his perspective of future research. He argued that the natural history of plants was such a demanding and extensive field that the life and diligence of one human, or even of several humans, was not sufficient to complete it (*ad perfectionem*). Only in the future would scholars extend and complete the existing knowledge by adding their own discoveries and observations to those of the present, until the discipline reached its highest perfection (Dodoens 1577, fols. *iiiv, v-r, v-v). Dodoens believed in future research; however, it was to consist of preserving and extending existing knowledge. Progress meant for him the further increment of knowledge until the totality of the unknown was exhausted and development was complete. The relativistic mechanism of the temporal regime of knowledge assumed that each extension of knowledge would manifest how much humans did not yet know because new findings provoked new questions. The ever elusive and expanding horizon of the human experience of nature was not consistent with the static image of nature as a theatre open to humans.[19] Therefore, the prescriptive idea of scientific progress is barely encountered in Renaissance natural history.

5.3 Metaphors of Unification of Experience

In the seventeenth century, natural philosophers discovered that nature was not a theatre offering itself to the human gaze; it hid inexhaustible riches of which the Greek philosophers had no idea. They further discovered that the scope of human

[18] Gessner 1552, unpaginated, quoted by Friedrich 1995, 84.

[19] Gessner 1551–1555, vol. 1, fol. α4v; Rondelet 1558, fol. a3; Bodin 1597, 633.

knowledge and experience was not unchangeable, but expanded as a result of exploration itself. This challenged the belief in the possibility of a compendious natural philosophy based on the limited amount of experience available to the individual. The study of nature thus called for the collective unification of individual experiences and their preservation over time.

At the same time, it was clear that the new collectivist concept of natural philosophy required a rethinking of the role of the individual. After explaining how Bacon and his followers in England dealt with Hippocrates's complaint about the brevity of human life, Yeo concluded his paper with an interesting remark: In the reflections of British natural philosophers on the progress of science, 'any agreement on how an individual life would contribute was difficult to achieve' (Yeo 2018, 93).

Yeo's observation about the lack of consensus among English natural philosophers can be explained by the absence of a concept of scientific progress that would be able to integrate individual life in a long history of knowledge. Such a concept seems to have been elaborated only later by Fontenelle and his contemporaries. Before Fontenelle, natural philosophers used a relatively settled canon of metaphors, similes, and tropes to articulate, in a nutshell, their ideas about the historical improvement of their field, including the disparity between the *vita brevis* and the *ars longa*.

Natural philosophers lacked original concepts and categories suitable for articulating their position in history therefore it was natural to them to find inspiration in Greek and Roman literature, which still largely shaped their thinking about the world, history and humans (Levitin 2015). A typical means of the conceptualization of history using classical sources was the Hippocratic aphorism *vita brevis, ars longa*, which runs throughout this chapter. Other devices include the metaphor *veritas filia temporis* and the simile of the dwarfs standing on the shoulders of giants. All of these served scholars to locate their natural philosophy in history and often implied a certain idea of the history of knowledge that sometimes approximated the idea of scientific progress but sometimes did not.[20]

I want to start this section with a trope that served primarily to deny the persuasiveness of Aristotelian philosophy and at the same time drew attention to the brevity of human life compared to the vastness of nature.

In his *Apologia pro Galileo* (1616/1622), which was intended to defend Galileo's telescopic discoveries against the objections of the Aristotelians, Tommaso Campanella wrote that nature was so vast that it could not be harnessed by the reason of a single man (Campanella 1622, 23). This trope was developed in various modifications by other representatives of early modern natural philosophy and mixed sciences, including Kepler, Galileo, Gassendi, and Power.[21] All of them were convinced that Aristotle's experience of nature was singular and brief,

---

[20] Duris 2016, 11–290; Blumenberg 1986, 153–172.

[21] Kepler 1938–2009, vol. 1, 16; Galilei 1890–1909, vol. 7, 136, 157; Galilei and Scheiner 2010, 128, 266, 291, 294, 357; Power 1664, fol. b2r; Fontenelle 1990–2001, vol. 2, 111; Fontenelle 1994, 51.

corresponding to his capacities and the duration of his life. However, natural philosophy based on individual experience could not account for the diversity of nature revealed by overseas discoveries, telescopes, microscopes, and experiments. At the same time, the principles of Aristotle's natural philosophy appeared to be refuted by rare phenomena that Aristotle could not have encountered during his limited life, such as new stars and comets. Glanvill articulated the trope simply: 'the advantages of Mankind in the Succession of two thousand years are more than those of a single person who lived but sixty-three.'[22]

Experience with the limitations of Aristotle's experience confirmed the insufficiency of any individual experience. As the individual in the short span of one life can only see a constrained empirical horizon, inductive conclusions based on individual experiences have very restricted validity. At the moment when the horizon of experience became mobile and the problem of induction appeared, a new bearer of knowledge had to emerge: humanity.

Humanity was linked to the development of natural knowledge by means of analogy between the development of humankind and the individual. This analogy, occurring in many versions, had a long intellectual history that early modern natural philosophers knew from classical ancient sources, from the Church Fathers, and from humanist literature.

The most influential version of this analogy was the rhetorical figure that W. von Leyden referred to as 'the reversed criterion of age'.[23] Renaissance philosophers considered the Greek philosophers wise and 'old', while they viewed the modern philosophers as 'young'. The figure of the reversed age was based on the assumption that wisdom resulted from rich experience. The old humans had more experience than the young ones; therefore they were wise. The simile carried this relationship of direct proportionality between age and wisdom to humanity and inverted the relationship between the Ancients and the Moderns in favour of the latter. The old and wise were not the humans from the beginning of history, but the last generation. If old and wise age was a manifestation of the passage of time, then current humanity must have been wise because it had the longest history. The figure thus allowed early modern philosophers to insist on the conjunction of old age and wisdom, and at the same time on a temporal epistemology according to which those who came last have the best knowledge.

We find the figure in Giordano Bruno and George Hakewill, but in both, it still appeared in the context of historical cycles in which periods of cultural flourish and decline alternated.[24] The specific perspective of early modern natural philosophy came only later, with Bacon who expressed the criterion of reversed age in one

[22]Glanvill 1668, 76; Glanvill 1885, 142; Wilkins 1802, 144; Hooke 1705, 360; cf. Dear 1985.

[23]Leyden 1958, 480. As far as I know, the trope was first noticed by Delvaille 1910, 184–188, 194–196. See also Guyer 1921; Schlobach 1980, 312–313; Armogathe 2001, 818–820; Taguieff 2002, 56.

[24]Bruno 1977, 96; Hakewill 1627, 23, 35.

economical sentence: 'And to speak truly, *Antiquitas seculi est Iuuentus Mundi*.'[25] The present age is old and more experienced, and may justly be expected to possess more perfect knowledge than that gathered by the young and inexperienced antiquity, as Bacon explains in his *Novum organum*:

> For the dotage and old age of the world should be taken for antiquity in its right sense, and ought to denote our times, and not the springtime of the world when the ancients lived. [...] And just as we expect greater knowledge of human affairs and maturity of judgments from an old man than from a youth [...] so in the same way much greater things could reasonably be expected from our time [...] than from the earliest ages, as is natural from the greater age of the world and from the countless experiments and observations which have been built up and stockpiled (Bacon 1996–2008, vol. 11, 133).

The reversal of the order of ancient and modern age eventually allowed Bacon to formulate a rhetorically compelling argument against Greek natural history, based on a new vision of his own position in history: Greek knowledge was the knowledge of children and as such had no relevance for the mature present age (Bacon 1996–2008, vol. 11, 114–115). The figure of the reversed age thus articulated the descriptive idea of progress by confirming the superiority of the later age. In this sense, the figure can be found in many seventeenth-century authors such as Wilkins, Comenius, Thomas Hobbes, Adrien Baillet, Glanvill, Cassini, and Malebranche.[26]

The analogy between the maturation of the human individual and the advancement of human knowledge, however, could also be developed in a more sophisticated form. It became a starting point for the metaphor of the 'universal man', in which the combination of wisdom, old age, and human history was translated into a developmental paradigm. The metaphor was based on the idea that the history of humanity was divided into successive phases corresponding to different ages of individual human life. Such an idea was already present in Christian thought of late antiquity. It related the history of humans to the history of the world and its gradual ageing, which would culminate in the Last Judgment.[27] In an eschatological sense, the idea also appeared in the early modern period. For example, Comenius, referring to Augustine, wrote, 'for the age of all mankind, is as it were the age of one man', which grew from childhood to adulthood through generations 'from Adam to the end of the world'.[28]

In the preface to *Tabulae Rudolphinae* (1627), Kepler conveyed the biological metaphor of the coming of age to the history of astronomy. Astronomy started 'from very slender and obscure beginnings'.[29] Later, like other living things, it 'gradually grew to adolescence'. Kepler described the course of astronomical history

---

[25] Bacon 1996–2008, vol. 4, 29. Bacon immediately explains: 'These times are the ancient times, when the world is ancient, & not those which we count antient Ordine retrogrado, by a computation backward from our selues'.

[26] Wilkins 1802, 138; Hobbes 1996, 474; Baillet 1691, vol. 2, 531; Glanvill 1885, 119; *Mémoires de l'Académie royale des sciences* 1729–1734, vol. 10, 56; Malebranche 1979, 213.

[27] Schmidt 1955–1956; Gatz 1967, 106–112; Dunphy 2010; Poulouin 2006, 32–33.

[28] Comenius 1651, 10; cf. Comenius 1964, vol. 2, 216.

[29] Kepler 1938–2009, vol. 10, 36; English translation Kepler 1972, 361.

connecting the different stages of the human age either with ancient peoples (the Greeks—*pueritia*) or with the great figures of the astronomical tradition (*adolescentia*—Hipparchus, *juventus*—Ptolemy). The last age mentioned is 'maturity' (*maturitas*), represented by Renaissance astronomy.[30]

Kepler's text should not be seen as expressing a prescriptive idea of scientific progress, for several reasons. Kepler's scheme could not account for the millennium-long medieval gap between antiquity and his time. Further, Kepler did not seem to have an elaborated vision of the further progress of astronomy, because it was not clear what should come after *maturitas*. Finally, in Kepler's concept, the course of history was still overseen by Divine Providence.[31] For Kepler, then, the maturation of astronomy was part of God's plan for history, not an expression of the idea of scientific progress occurring independently of the history of the world and the plans of Providence.

Early modern philosophers took an innovative approach to the old developmental metaphor of the ages of humanity. They abandoned the eschatological framework of the idea and concentrated only on the epistemological side. They imagined the history of natural philosophy as a development of man that unified all human experiences and was gradually learned during development from childhood to adulthood.

In his *Préface sur le Traité du Vide*, Pascal first introduced this new bearer of knowledge. According to Pascal, the starting point for the consideration of progress was the idea of a caring nature that chose a different approach to animals and humans. Nature provided animals with the instinctive or innate knowledge that helped them survive without having to learn anything. Living creatures, however, were neither able to extend nor transmit this innate knowledge. Animal knowledge was born and lost with each individual and was therefore immutable. It was not the same with man, 'who is formed only for infinity' (*qui ne produit que pour l'infinité*).[32]

The infinite in Pascal's text did not have an existential or cosmological meaning. It concerned knowledge. Humans were not endowed with any instincts or innate knowledge, but they could learn and pass on knowledge. In a limitless world, moreover, they could expect further expansion of knowledge resulting from the inexhaustible productivity of the senses and reason, which could create new tools and methods of knowledge.

Pascal argued that at the beginning of life, the human individual lived in a state of ignorance, 'but he is instructed unceasingly in his progress'. Pascal's individual used his own experience and also the knowledge of his predecessors. 'And as he preserves this knowledge, he can also add to it easily.' Thanks to this 'special privilege'

---

[30]Kepler 1938–2009, vol. 10, 37–39; cf. Grafton 1992, 202.

[31]Kepler 1938–2009, vol. 1, 329–332; vol. 1, 6; vol. 6, 290. See also Kepler's letter to Herwart, April 1599, Kepler 1938–2009, vol. 13, 309: *Haec sunt intra captum judicii humani, haec nos scire deus voluit, dum ad suam nos imaginem condidit.*

[32]Pascal 1998, 456; similarly, Bossuet 1881, 198.

humans advanced from day to day in the sciences, and all mankind progressed together 'in proportion as the world grows older'. In the succession of human beings, the same thing was repeated as in the succession of the different ages of the individual. Therefore, generations of humans with all their accumulated experience were to be considered 'universal man' (*l'homme universel*), 'who subsists forever and learns continually'.[33] The man, 'who is formed only for infinity,' is the generic man, a sum of all experiences of humanity.

Humans were not formed for infinity as individuals, but as a genus—as a unified subject of history that persisted for centuries, constantly learning and sustained by human effort. The unfinished universal man, who continually extended the boundaries of his knowledge, corresponded to the unfinished universe, which changed and aged. The absence of anthropocentric teleology was compensated for by Pascal through the idea of scientific progress. Knowledge was able to keep growing by means of the continuity of the genus, the growth of experience, the rational method, and human diligence.

The idea of the eternally learning man was naturally associated with the figure of the reversed age, which legitimized the break with classical antiquity. According to Pascal, the Ancients 'constituted the infancy of mankind'. The wisdom of old age was reserved for the Moderns, who stood at the end of history and therefore had more experience.[34] The main aim of Pascal's text was probably to reject reverence for antiquity in the context of air-pump experiments (Mazauric 1998, 106–110), yet his conception of the great man represented a relatively sophisticated conceptualization of the historical unification of human experience which, in addition to justifying its superiority over antiquity, opened up the prospect of further progress.

The metaphor of the universal man reappeared thirty years later in the context of the quarrel of the Ancients and the Moderns. In the first volume of *Parallèle des Anciens et des Modernes* (1688), Charles Perrault argued that the duration of the world could be understood as the life of a man who passed through childhood, youth, and adulthood and who was now in old age. Human nature was also such a person who experienced childhood during the infancy of the world. Youth and mature age followed. Eventually, the world and man had reached old age. Naturally, the reversed criterion of age also appeared in Perrault, who wrote that our first parents should be regarded as children, while the present generation 'should be regarded as old men and true ancients of the world' (Perrault 1688–1697, vol. 1, 34).

At the same time and in the same intellectual debate, Fontenelle also used the metaphor of the universal man. In his *Digression sur les Anciens et Modernes* (1688), Fontenelle wrote that each educated spirit was composed of all the spirits of preceding centuries, and thus only one spirit had been educated during all that time. The history of human knowledge was the history of the education of a single man. This man lived from the beginning of the world, had his childhood and his youth, and had now reached 'the prime of life' (*l'âge d virilité*) when he was most

[33]Pascal 1998, 456; English translation in Pascal 1910, 449.

[34]Pascal 1998, 456; English translation in Pascal 1910, 449.

reasonable and most enlightened. Fontenelle concluded his simile with the following remark: 'It is a shame to carry to the end a comparison which is going so well; but I am forced to admit that mankind will never have an old age.'[35]

Fontenelle thus refused to develop the metaphor further to the senescence of universal man, which would imply intellectual flaccidity and decline. Instead, he insisted that the next centuries would be as productive as the young age. 'That mankind will *never* degenerate, and the sound views of all subsequent thinkers will *forever* be added to the existing stock'.[36] Therefore, neither decline nor extinction could be expected in the future, but only steady growth.

As shown by Rouvillois (2010, 163) the metaphor of the universal man can also be found in the later *Observations sur le Progréz Continuel* (1737), written by Fontenelle's friend and roommate, Abbé Charles Irénée Castel de Saint-Pierre. Saint-Pierre also identified the phases of history with the human ages, of which a single subject was the bearer. He stressed, however, that humanity, though nearly 8000 years old, 'is only in the infancy of human reason'. Compared with what was to come in another 5000 or 6000 years, the humankind stood at the beginning of early youth (Castel de Saint-Pierre 1737, 274–275).

In his inspiring analysis of these well-known texts, Rouvillois (2010, 163–164) emphasized an important fact: the cited authors had not assigned universal man the same age. Pascal and Perrault saw in contemporary humanity an old man, Fontenelle attributed to it a manly age, Saint-Pierre a childhood. This age difference, according to Rouvillois, lay in the different functions that the metaphor had for each author. Pascal's and Perrault's old man symbolized actual humanity in relation to their predecessors, based on a comparison of the amount of knowledge. The predecessors knew less because they had not had enough time to accumulate a sufficient amount of knowledge. Contemporaries knew more because they could rely on data accumulated for a longer period. Senescence symbolizing wisdom was a picture of the totality of the experience of existing humanity, which had been gradually added to (Rouvillois 2010, 164).

The notion of continuous ageing and accumulation of knowledge, however, implied the impossibility of further developing the metaphor: The metaphorical old age of the Moderns provided no perspective. Pascal did not further develop the fate of the universal man and confined himself to metaphysical statements about the infinite without making the idea of future expansion of knowledge more precise. Perrault thought that the old age of universal man heralded the culmination of the ascending phase of the historical cycle and the approaching time of cultural decline (Perrault 1688–1697, vol. 1, 67). Fontenelle recommended abandoning the metaphor at the moment when one started talking about the future. Despite the manly age, the whole process of educating humanity would continue 'forever' and would be filled

[35] Fontenelle 1990–2001, vol. 2, 426; English translation in Fontenelle 1960, 366–367 (emphasis mine).

[36] Fontenelle 1990–2001, vol. 2, 426; English translation in Fontenelle 1960, 367 (emphasis mine).

with 'that endlessly growing stock of ideas'.[37] Saint-Pierre found a solution in the relativization of the metaphor. His universal man was old in relation to the past, but a child in relation to the future.

The differences in the understanding of the ages of universal man show a fundamental reconfiguration in the understanding of the progress of knowledge. Originally, natural philosophers used the descriptive idea of progress to legitimize the new natural philosophy by highlighting its superiority over Aristotelian natural philosophy. By the time of the emerging Enlightenment, Saint-Pierre took the superiority of the Moderns for granted and felt no need to address it. Of central importance to him was the prescriptive aspect of the idea of progress, which, although it included a comparative component, manifested itself above all as an expectation of future achievements—as an awareness that humanity had only just begun the great work. Based on the same belief associating wisdom with old age, the descriptive idea of progress metaphorically made contemporary humans old. The prescriptive made them children who could expect the culmination of knowledge only in the future.

Fontenelle's man of manly age, who continued to age and yet developed forever, stood somewhere between these positions. Hence the whole metaphor was, as Sophie Audidière noted, more obscuring than illuminating (Audidière 2016, 28). In fact, it turned out that Fontenelle had better conceptual means of expressing the gathering of human experience—the idea of the progress of the human spirit.

5.4 Collaboration Over Generations

Let us now turn to the different way of overcoming discrepancy between *vita brevis* and *ars longa*. This approach methodologically unified the experiences of philosophers and scientists working in succession in history. In this case, individuals had to concede to deliberately handing over the results of their work to the future, where the final synthesis would be made. The origins of this strategy can be found in Bacon.

In his *Instauratio magna* (1620), Bacon laid out his project for a great renewal of the sciences in six parts, intended to imitate the six days of creation. He believed that the project would one day be completed with a seventh part when the sabbath of the sciences would occur: a time of rest and completion (Bacon 1996–2008, vol. 11, 47). Pursuing the great instauration of sciences, humans should imitate God's act of creation and work toward the day when labour and toil would cease, for they would become masters of nature. In *De Augmentis Scientiarum* (1623), Bacon expressed the finality of the work of restoring the sciences in another way. Humans should build knowledge as a pyramid, with natural history as its base and natural philosophy, physics, and metaphysics as the next levels. Bacon used the Hippocratic saying *vita brevis, ars longa* to articulate his view that working on the pyramid of the

[37] Fontenelle 1990–2001, vol. 2, 426; English translation in Fontenelle 1960, 367.

sciences went beyond an individual life (Bacon 1857–1874, vol. 1, 657; cf. vol. 3, 227).

Bacon was aware that he instauration of sciences required the accumulation of a vast amount of empirical findings in many natural histories. Bacon did not hope to accomplish such a task in his lifetime. Completion of the work 'is not confined entirely to a single age but to a succession of them' (Bacon 1996–2008, vol. 11, 25). Bacon saw himself as a thinker who did not bring definitive conclusions but only prepared material for future achievements that he himself would not live to see.[38] Bacon's philosophical project thus added to the tradition of searching for the *vita longa* a perspective in which the individual relinquished the completion of knowledge in his lifetime and passed on an unfinished and incomplete work to posterity.

Descartes thought differently about the future of natural philosophy. Even more resolutely than Bacon, he insisted that he owed nothing to the previous philosophical tradition. Descartes saw his philosophy not as the result of slow previous progress, but as a sudden break in the history of error and a new beginning of knowledge (Kamlah 1961; Cottingham 1992). Descartes was also convinced that he owed nothing to contemporary philosophers. Working entirely independently and having almost no contact with anyone, he insisted that he had derived the correct method and the true principles of knowledge and nature himself. He also believed that he alone could best process and organize all knowledge.[39] Finally, he expected himself to conclude the whole system of natural philosophy by formulating definitive wisdom.

According to Mazauric, in Descartes's philosophy, it was necessary to distinguish between two kinds of truths with respect to time. The first kind were truths obtained by deduction, which were more or less independent of time. The second kind were empirical truths, which required time and patience. Deductive truths gave hope for an early conclusion in the form of a complete system of knowledge. Empirical truths tempered this hope because Descartes was well aware of the extent of empirical nature and knew that it was impossible to deduce the whole of nature from general principles.[40] Therefore, according to Mazauric, Descartes always wavered between two perspectives: the early conclusion of the sciences and the postponement of their completion to the future (Mazauric 2007, 254–259).

Both options can be found in the introduction to the French translation of the *Principia philosophiae*. Here Descartes implied that he did not have enough resources to conduct all the experiments, and therefore the completion of the whole work would come only in the future: 'our descendants may see the happy outcome of this project' (*en voient le succez*).[41] In the same text, however, Descartes wrote that he did not yet feel so old, or so diffident about his powers. He hoped he

[38] Bacon 1996–2008, vol. 11, 44–45, 196; cf. vol. 13, 260; cf. Bacon 1857–1874, vol. 1, 492–493.

[39] Dear 2015, 2015, 78; cf. Schouls 1987; Werle 2007, 411–412.

[40] Descartes 1996, vol. 6, 64; similarly vol. 9–2, 20; cf. Clarke 1982, 143–144; Clarke 1992.

[41] Descartes 1996, vol. 9–2, 17, 20; English translation in Descartes 1984–1991, vol. 1, 190.

could 'boldly try to bring the plan to its conclusion'.[42] He felt capable of completing his project and finishing it in wisdom (*sagesse*), which was the 'perfect knowledge of all things that mankind is capable of knowing'.[43]

Descartes's hope for success in his own lifetime came from several sources. He cherished the hope for a medical extension of life (Shapin 2000). Further, his definition of wisdom suggested that the reservoir of possible knowledge was exhaustible, and therefore the advancement of natural philosophy would be terminable. Another source of his self-confidence was the conviction that his method would accelerate the process of expanding knowledge. Descartes also believed that if he could obtain wealthy patrons, he could employ assistants to carry out his instructions (Descartes 1996, vol. 6, 63, 72, vol. 9–2, 20). Finally, he expected other philosophers to send him their raw data for further processing (Descartes 1996, vol. 6, 65).

Similarly to the Renaissance scholars, Descartes believed in the individual intensification of research through hired help, money, his own effort, correct methods, and the willingness of others to accept his guidance. The idea of institutional, collective cooperation on a shared task was alien to Descartes. He embodied the figure of the private, independent scholar who sought primarily a wealthy patron, not collaborators for a team.

Descartes also grossly underestimated the role of time for knowledge. His natural philosophy owed supposedly nothing to its predecessors, and the truths he discovered were expected to last. Descartes remained in an epistemologically constrained world that could be described by a finite number of truths and principles that were already essentially complete. He intended to pass on his principles of philosophy as a gift to posterity, which he saw as receiving a great work, 'a body of philosophy that is quite complete'.[44] Posterity was to complete the work by filling in the gaps in the accomplished theoretical structure with the missing empirical data. Descartes's philosophy was not meant to accept the temporal regime of knowledge in which solved questions opened up completely unexpected problems.

Thus Descartes became the founder of a school that, while asserting its own superiority over classical antiquity and advocating the descriptive idea of progress (Rohault 1671, fols. a3r–a4v), did not itself make any real progress. Cartesians never accepted the temporal regime of knowledge because they insisted on the rigid structure of the principles discovered by their founder (Clarke 1989, 222, 238–240).

Other early modern natural philosophers accepted the idea that natural philosophy and mixed sciences had to escape the boundaries of an individual life. They expressed their beliefs using the similes and metaphors mentioned above or drew on classical literature and other inspirations to conceptualize their own position in the history of knowledge. This brings us back to Seneca.

---

[42] Descartes 1996, vol. 9–2, 17; English translation in Descartes 1984–1991, vol. 1, 188.

[43] Descartes 1996, vol. 9–2, 2; English translation in Descartes 1984–1991, vol. 1, 179.

[44] Descartes 1996, vol. 9–2, 2, 17; English translation in Descartes 1984–1991, vol. 1, 188.

In his *Naturales Quaestiones,* Seneca made the famous statement about descendants who would know comets better than his contemporaries did. Two other important passages in the same work played a role in shaping the early modern idea of scientific progress. In the first passage, Seneca introduced the classical idea of personified nature, which gradually revealed itself to humans and initiated them into its secrets. Here Seneca stressed that many things were not yet known to the people of his age: 'The people of a future age will know much that is unknown to us; much is being kept for the generations to come after the memory of us has faded away. The world is a paltry thing unless it contains something for every age to discover.' To this Seneca added that knowledge of the whole world was still very far off: 'Great enterprises progress slowly, especially if effort falters (*si labor cessat*).'[45]

Seneca here formulated an idea that could be called the principle of the even distribution of discoveries in history. In contrast to Aristotelian teleological progressivism, this principle envisaged the incompleteness of knowledge and the possibility of its further improvement. Each generation and each age had reserved the right to its own discoveries, and therefore progress in the knowledge of the whole world would proceed slowly, following the succession of human generations.

Elsewhere in *Naturales Quaestiones,* Seneca added a developmental perspective to the principle of the even distribution of discoveries that denied the possibility of perfect knowledge at the beginning of history:

> Nothing is completed while it is beginning [. . .] Even though much will have been done on this subject every age will none the less find something to do. As in every other subject, the first beginnings have always been far away from the completed knowledge (Seneca 2014, 92).

In view of some scholars, these words may indicate that Seneca anticipated the modern idea of progress in *Naturales Quaestiones* (Motto 1984). I argue that Seneca's statements were only occasional remarks that did not result in any research programme or important discoveries in classical antiquity. This was particularly striking in comparison with astronomy, where the patient collection and transmission of data over successive generations became a program, the commonly shared methodology of the field.

Early modern natural philosophers often quoted Seneca's statements and referred to them directly and indirectly. Using the words of the classical philosopher, they could present a new form of acquiring knowledge. In his astronomical treatise *Mercurius in Sole Visus* (1631), Gassendi repeated Seneca's words about future generations who would wonder that their predecessors did not know such obvious things. Gassendi added that posterity would know more, just as the contemporary philosophers knew more than their predecessors had known.[46] In his 1654 biography of Tycho, Gassendi wondered why humans had not been able to achieve more perfect astronomical knowledge before his time. As an answer, Gassendi used the

[45] Seneca 2014, 134; Williams 2012, 265–270; Hadot 2006, 166–181.

[46] Gassendi 1658, vol. 4, 504a, cf. vol. 6, 209, vol. 1, 82a.

quotation from Seneca about the first beginnings to describe an implicit historical framework that assumed a gradual development from ignorance to advanced knowledge. The authority of the Roman philosopher was meant to reassure contemporaries that the limitations of human knowledge were neither mysterious nor unnatural, but the result of a slow development that was an intrinsic part of any human inquiry.

In his *Machina Coelestis* (1673), Hevelius also referred to Seneca. Although much had been discovered, Hevelius explained, even more remained for future astronomers. Every age has something to do, and it cannot be assumed that we have exhausted all of nature (Hevelius 1673, vol. 1, 26). Similar direct and indirect references to Seneca can be found in Wilkins, Pascal, Glanvill, Huygens, Boyle, Cassini, or Clarke's notes to Rohault's overview of Cartesian natural philosophy.[47]

Early modern natural philosophers subscribed to the principle of the even distribution of discoveries for several reasons. First, there was a respectable literary model for its expression. Further, they discovered a majestic range of empirical nature that was not adapted to human abilities. They also embraced the relativistic mechanism of progress that turned every piece of knowledge they attained into a link in an unfinished chain.

In his *Entretiens*, Fontenelle also admitted the even distribution of discoveries: 'Do we presume to have discovered all things, or to have taken them to the point where we can add nothing? For goodness' sake, let's admit that there'll still be something left for future centuries to do.'[48] Fontenelle did not quote Seneca, as he might not have needed to legitimize his conviction of further progress by alluding to an ancient text. He derived his belief in the principle of the even distribution of discovery from the context of *Entretiens*, where he wrote about the vastness of nature, the limitations of human cognitive faculties, and the temporal regime of knowledge.

Insisting on the validity of the principle of equal distribution of discoveries, natural philosophers were forced to admit that they would not be able to conclude knowledge in their lifetime. No individual subject was capable of unifying all experiences and formulate definite conclusions. This view significantly changed the way the natural philosophers thought about the role of descendants. Renaissance authors such as Tycho and Gessner left their work to posterity as a more or less closed legacy. The descendants were the recipients from whom admiration and gratitude were expected. For the natural philosophers of the second half of the seventeenth century, subsequent generations were no longer to be passive recipients but collaborating colleagues.

In his *Plus Ultra,* Glanvill said that the telescopes had shown us 'immense Territories of Light', so it was correct to anticipate the future generations making further discoveries: 'And the present Philosophers are so far from desiring that

[47] Wilkins 1802, 110; Pascal 1998, vol. 1, 453; Glanvill 1668, 64; Huygens 1888–1950, vol. 19, 455, vol. 21, 357; Boyle 1775, vol. 2, 35; *Mémoires de l'Académie royale des sciences* 1729–1734, vol. 8, 18; Rohault and Clarke 1735, 85.

[48] Fontenelle 1990–2001, vol. 2, 54; English translation in Fontenelle 1990, 34.

Posterity should sit down with their Discoveries and Hypotheses...' (Glanvill 1668, 55). Similarly, Sprat stressed that fellows of the Royal Society were merely collecting and sorting facts and passing them on to the future. 'By this they have made a firm Confederacy, between their own present Labours, and the Industry of future Ages.' If the same alliance had been made with us by the Greeks, Sprat complained, 'what Depth of Nature could by This Time have been hidden from our View?'[49] *Vita longa* was thus achieved by drawing the descendants into the process of knowledge as active participants, temporary unknown members of the research team.

This radical opening of knowledge towards the future and the anticipation of colleagues from subsequent generations can also be found in French natural philosophy. In his posthumously published *Syntagma Philosophicum*, Gassendi argued that many things, such as mites or stars, remained hidden from the ancient philosophers but were obvious to contemporary philosophers (Gassendi 1658, vol. 1, 82a). Gassendi did not limited himself to pronounce the superiority of modern natural philosophy over the Ancients. He transcended the descriptive idea of progress by extending knowledge to the future. In the spirit of the temporal regime of knowledge, he held the suspicion that many things were hidden from the present generation. There was plenty of unexpected work to do for philosophers of the future. Descendants were therefore not mere recipients of gifts and bequests, but active completers of knowledge (Gassendi 1658, vol. 1, 82a).

The establishment of a 'confederacy' with the philosophers who were yet to come meant that the current generation was to provide the descendants with a disposition for future success. That is why early modern philosophers started to understand their own historical position as being in service to the future. This self-understanding followed from the basic tenets of the temporal regime of knowledge such as a rejection of the principle of authority, a refusal to formulate conclusions dogmatically, an aversion to the hasty creation of speculative systems, and openness to evidence challenging established theories. This constraining self-discipline resulted in the implementation of a Baconian programme of collecting and transmitting experiential data. Future generations were to be involved in research by having the current generation prepare empirical data for them. Glanvill summarized this position of the English Baconians in a few words: 'We must seek and gather, observe and examine, and lay up in Bank for the Ages that come after' (Glanvill 1668, 91–92).

On the continent, Gassendi introduced his readers to a similar notion of engaging future generations when he reflected on the epistemological problems of astronomy. Above all, Gassendi stressed the importance of making and transmitting observations. Theoretical endeavours did not play a major role for him in the history of astronomy: 'Observations are the foundations on which astronomy is built' (Gassendi 1658, vol. 6, 46b). Unlike theoretical considerations, observational data retain their value in time. Gassendi was aware that astronomers sometimes projected their own views into observational data, and he therefore emphasized the practice of

[49] Sprat 1734, 116; Hooke 1705, 21–24.

producing lists of 'naked observations' (*nudas observationes*) that were not burdened by theoretical speculation. These observations were to be passed on to future generations who would evaluate them and form their own hypotheses.[50] Future natural philosophers would combine, evaluate, systemize, and compare the data. Eventually, they would insert the inherited data into a—at the time unimaginable—theoretical framework (Gassendi 1658, vol. 5, 474b).

Gassendi's empirical methodology demonstrably shaped the views of the founding figures of the Académie, such as Du Hamel, Picard, Auzout, and Roberval (Niderst 1972, 121). The representatives of the empirical sciences further developed empirical epistemic practices. Especially the astronomers from the Observatoire, focused on team accumulation of data for later use which Daston has called collective empiricism (Daston 1995, 2011). According to this practice, the observations had to be utilizable, collective, coordinated, long-term, and free of errors.

At the simplest level, collective empiricism was practiced by astronomers making joint observations when several academicians observed the same phenomenon from the same location in Paris with different instruments.[51] Further, they developed a procedure that Daston called 'coordinated observation' (Daston 2011, 87–91). This epistemic practice was based on the fact that the determination of certain geographical or astronomical values required the observation of the same phenomenon from several locations on the Earth's surface at the same time. In the 1670s and 1680s, the astronomers of the Académie regularly made coordinated observations from various locations in France.[52]

With Colbert's support, astronomers were able to greatly expand the observational base for coordinated observations by arranging the first scientific expeditions. In 1670–1671, Jean Picard travelled to the Danish island of Hven, where he made observations in the ruins of Tycho's observatory. In 1672–73, Jean Richer stayed in Cayenne, a French colony in South America. Picard and Richer made a series of observations in coordination with astronomers who were observing the same phenomenon simultaneously in Paris and elsewhere in France. Both expeditions proved to be a great success because the comparison of the measured data led to significant improvements in some astronomical parameters, values, and models.[53] Most importantly, however, it showed that institutionally organized, coordinated observations

[50] Gassendi 1658, vol. 4, 75; vol. 6, 29a-b, cf. also letter to Wendelin, where Gassendi warns against accommodating observations to preconceived hypotheses, vol. 6, 209a.

[51] Huygens 1888–1950, vol. 6, 59–67, vol. 15, 118, 120, 125–128, 483, 498; Fontenelle and Godin 1733, vol. 1, 9; *Mémoires de l'Académie royale des sciences* 1729–1734, vol. 10, 478; Cassini 1673, 12–14.

[52] See shorts tracts comprising various observations made by Picard and La Hire in *Mémoires de l'Académie royale des sciences* 1729–1734, vol. 7, 358–429, here 349, 351, 379, 399, 413; see also many short reports in vol. 10, 151–153; 571–572, 602, 605, 612, 621; see also Fontenelle and Godin 1733, vol. 1, 181, 206, 212, 264, 331, 411.

[53] Cassini summarized the achievements of expeditions in Cassini 1684, 45–49; cf. Olmsted 1942; Van Helden 1989, 118.

provided results of precision and quality that had been unavailable to generations of previous astronomers.

An important feature of collective empiricism was that observations were made systematically over a long period of time. Cassini was the perfect embodiment of this epistemic practice because he made regular observations of the same phenomena over decades. For instance, he observed spots on the surface of Jupiter continuously for 40 years (Fontenelle and Godin 1733, vol. 1, 106–108). The author of the second volume of the history of the Académie, Louis Godin, recalled in this connection that Cassini made and recorded these observations precisely to make them useful for posterity. Cassini hoped that his systematic, long-term, and accurate observations would be much more useful to future astronomers than the imprecise observations of classical antiquity that he had at his disposal (Fontenelle and Godin 1733, vol. 1, 104). This remark suggested that Cassini was disappointed by the range and accuracy of the past observations that had survived and hoped to provide future astronomers with the tools he lacked.

Similar considerations about future research can be found outside of astronomy. In *Mémoires pour servir à l'histoire des planets*, Dodart enumerated a list of tasks to be undertaken by natural historians in the future. He and his team added to and perfected the Greek and Roman works, but they did not manage to complete the whole of natural history, a task for future researchers (Dodart 1679, 7, 160, 305). Involving future scientists in research was therefore no longer just about passing on data, but also about formulating research questions, tasks, and problems that could not be answered at the time.

No matter how large the reservoir of empirical data was, academicians expected the exploration of nature to continue in the distant future. Even though they had observations collected for decades, it was still not enough. In the *HARS* volume for the academic year 1699, Cassini gave an account of observations of comets in recent decades that he or other astronomers had made. Yet he insisted that he did not have enough accurate observations over a long enough period of time to formulate a theory predicting the returns of comets. Cassini said that sciences of his time were confined to describing and comparing comets. Only in the future, after sufficient amount of observations, would it be possible to determine inductively the patterns of their motion (*Mémoires de mathématique et de physique* 1702–1742, vol. 1, 41, 43). When Fontenelle summarized this text by Cassini in the same volume of *HARS*, he also emphasized that astronomers would have to wait several centuries to discover these regularities (Fontenelle 1702–1742, vol. 1, 75).

Because of the longevity of the astronomical agenda, complaints about the shortness of life are repeatedly encountered in Cassini's texts. Cassini was trying to discover a possible connection between the appearance of spots on the surface of Jupiter and the revolution of the planet (12 years). Exploring this subject, he bitterly noted that these observations 'teach us that there are changes in nature, of which we can only become aware after a number of years, sometimes greater than is granted to the life of one man' (*Mémoires de mathématique et de physique* 1702–1742, vol. 1, 105). In another text from 1692, Cassini reported his observations of Jupiter in opposition to the Sun, a position that provided a suitable opportunity to observe spots on its surface. Cassini formulated several conclusions, but he concluded the report by saying that it would be necessary to wait for another similar opposition of Jupiter with the Sun to verify his conclusions. Such an observation would remain reserved for our grandchildren, Cassini complained, because the next opposition of Jupiter and the Sun at the same degree of the ecliptic would return only in 83 years (*Mémoires de l'Académie royale des sciences* 1729–1734, vol. 10, 9).

Simultaneous and coordinated observations from various locations by various observers were possible to operate meaningfully only due to the standardization of observational practices. Cassini repeatedly pointed out that differences in observational practices led to errors. Measurements always had to be made in the same way (Cassini 1681, 33, 80). Cassini, therefore, developed detailed instructions for his colleagues at the Observatoire and other observers on how to proceed with observations (*Mémoires de l'Académie royale des sciences* 1729–1734, vol. 7, 432–438). Every observational task, the setup of the instruments, the recording of the results, and the course of the whole event were to take place in the same way, if the comparison of the data was to yield valuable results. The observational practices had to be standardized and turned into routine (Špelda 2022).

Reading Cassini's instructions makes clear that he wished for a depersonalization of experience. The collective experience accumulated during the *vita longa* was to be purged of all individual features. Collective empiricism demanded not only the avoidance of hasty speculation but also the mechanization of cognitive habits and procedures. To make good observations, the astronomer had to have a trained eye, a sure hand, automated habits, and mechanical reactions that could only be achieved by constant repetition. The mechanical symbiosis between humans and their instruments was to overshadow the personality of the astronomer.

The same remark that Harrison formulated for early modern natural philosophers applied to astronomers:

> Central to the prestige of scientific methods is their insensitivity to the personal qualities of those who employ them. The putative universality and objectivity of science are attributed to the fact that the production of dependable knowledge does not rely on its practitioners sharing a common set of personal characteristics, but rather on their observance of a common set of procedures (Harrison 2006, 220).

Astronomy and experimental research did not require a brilliant thinker deciphering the secrets of the cosmos, but a patient observer who sat down every evening at their instruments to repeat the same observations and measurements over and over again. This form of research required dedication to the field, faith in the institution, and professionalism. The natural philosopher and scientist turned into an employee of an institution that was both a guarantor and a subject of knowledge.

5.5 Institutional Immortality

On 4 February 1699, Fontenelle read the members of the Académie a document giving the institution new organizational principles. In collaboration with Fontenelle and Minister Pontchartrain, Abbé Bignon, president of the Académie, devised new rules for the functioning of the Académie, the *règlement*.[54] It was such a major innovation that Fontenelle labelled it the 'second birth' of the Académie.[55] Using the history of the Académie I want to present a final means of resolving the disparity between short human life and vastness of nature: institutionalisation of natural philosophy and sciences.

Before 1699, the Académie had followed almost no rules except for compulsory attendance at the regular twice-weekly meetings. No obligations of the state towards the academicians were established, leaving them at the mercy of the discretion of the ministers who were responsible for the Académie. During his reign, Colbert left academicians to set their own research tasks. Indeed, according to Sturdy, Colbert optimistically hoped that by 'letting the academicians run their own affairs', they would meet his utilitarian expectations (Sturdy 1995, 77). This self-regulating system was relatively functional during Colbert's reign, but it broke down under Colbert's successors, and the institution had to be reformed.

Previous academies, such as the Montmor Academy in Paris, were ephemeral and died out with their patrons. After Colbert's death, contemporaries expected the institution to suffer the same fate as those previous academies (Dew 2015, 431). However, this did not happen. In 1683, the Marquis of Louvois took over the charge of the Académie; after his death in 1691, the Comte of Pontchartrain became a new ministerial protector of the institution. The new minister, together with Abbé Bignon, had apparently been thinking of a closer relationship between the Académie and the monarchy. The *règlement* of 1699 then explicitly declared the Académie to be a royal facility. Thus, according to Hahn, Stroup, and Mazauric, after the reform, the Académie ceased to be a matter of personal decisions and became a public institution that was part of the state apparatus, which imposed its goals, duties, and research orientation.[56]

The *règlement* of 1699 contained 50 articles that divided academicians into four groups with specific tasks and duties.[57] The state's obligations to academicians, especially regular financial support, were spelt out. Individual articles governed the day-to-day running of the academy. Academicians were obliged to attend meetings

[54] The document was originally printed as 'Préface' in Fontenelle 1702–1742, vol. 1 (1699), 1–13. In Fontenelle's *Œuvres complètes*, the text was published as 'Histoire du renouvellement de l'Académie royale des sciences', see Fontenelle 1990–2001, vol. 6, 53–69. English translation is available in Fontenelle 1727, xviii–xxix.

[55] Fontenelle 1990–2001, vol. 6, 53; cf. Blechet 1989.

[56] Hahn 1971, 76; Mazauric 2007, 152; Stroup 1990, 37–38.

[57] Hahn 1971, 58–83; Niderst 1972, 190–191, 201–208; Sturdy 1995, 281–295; Tits-Dieuaide 1998; Poulouin 2006; Mazauric 2007, 58–71; Shank 2018, 217–224; Seguin 2022, 2023.

and to reside in Paris. Members of the institution were required to repeat and confirm experiments made by other researchers in other countries. They were also obliged to submit reports on their work and to present plans for the future. Another duty of the Académie members was to submit documents to the secretary with the comments they made during the discussions at the meetings. Publications by individual members had to be approved by the body of the Académie. The duties of the academicians also included assessing the inventions and machines that their inventors wished to patent.

Historians agree that the institutional transformation of 1699 brought about the professionalization of natural philosophy. This meant several things. Natural philosophy and the mixed sciences ceased to be the intellectual pastimes of private scholars and turned into paid employment with opportunities for career and social advancement (Ruellet 2016, 81–83, 119). In the early Académie days, remuneration was designed for personalities: 'pensions were assigned directly to individuals and not to the position of academician per se' (McClellan III 1981, 543). After the reform, academicians received pensions for complying with obligations according to their rank in the institutional hierarchy.[58] The monarch no longer supported individuals but the research itself and the advancement of individual disciplines. This change also meant that the original informal freedom of research and choice of interests was balanced by normalization, standardization, and adherence to epistemic practices and performance requirements.

Some articles of the *règlement* also regulated the rules of academic conduct. John B. Shank explained the ethical aspects of the behaviour of academicians and some epistemological attitudes (sage Pyrrhonism) as resulting from the influence of the social norms of French society at the time—such as the entry of gentlemanly behaviour (*honnête homme*) into epistemology and scientific research (2018, 303, 319, 340–341). Unlike Shank, I believe that at least some of the articles were much more likely to express an experience with the temporal regime of knowledge that had been successfully cultivated at the Académie since 1666. For example, Article XXVI required academicians to recognize the legitimacy of other views, to be prepared to compromise, to be willing to change their minds, especially in the face of empirical evidence, and to avoid expressing a definitive position on a scientific problem (Fontenelle 1990–2001, vol. 6, 60).

While the article can be interpreted as regulating academic debate through the ideal of the *honnête homme*, in my view the requirements of Article XXVI were a natural reflection of the epistemology of sage Pyrrhonism, which demanded openness and giving up obtaining definitive answers. Members of the new scientific institution were to control their professional conduct so that the institution could make further progress. Epistemology thus helped to create professional ethics. This was evidenced by the fact that in the *éloges* Fontenelle was also able to transform the epistemological rules of the temporal regime of knowledge into epistemic virtues

[58] Hahn 1971, 51, Stroup 1990, 223–224; Mazauric 2007, 18, 64–66, 106–108; Grimaldi 2020, 28–29.

such as intellectual humility, patience, and open-mindedness. For example, Fontenelle stated that the mathematician l'Hôpital was ready to admit 'his Ignorance and to receive instructions' (Fontenelle 1990–2001, vol. 6, 106). Cassini, according to Fontenelle, communicated his discoveries not out of vain boasting. His knowledge should serve 'to the progress of science, not to his personal glory' (Fontenelle 1990–2001, vol. 6, 291).

Article XX stated that experience had shown that the collective work of the whole Académie had produced 'many Inconveniencies'. In view of the authors of the reform, academicians should further concentrate on their own specific tasks, informing others and enjoying their comments.[59] Marie-Jeanne Tits-Dieuaide explained the significance of the article by noting that in the early Académie, following the Baconian spirit, all academicians participated in everything. For example, mathematicians had to take part in dissections, which Huygens disliked.[60] Tits-Dieuaide noted that Baconian collectivism was a 'simplistic and utopian conception of research' that matched neither the inclinations nor the talents of individual scientists who were unwilling to sacrifice their reputations and prestige on the altar of science (Tits-Dieuaide 1998, 102–103).

The article thus sought to alleviate the tension between the ideal of service to a transpersonal subject and the *vita brevis* of particular researchers who continued to strive for reputation and were not willing to dissolve their individual goals and interests in the collective body of the institution. Historians have repeatedly emphasized that early modern natural philosophy required a new type of *persona* compared to the scholastic and Renaissance natural philosophy (Gaukroger 2006). The collective, state-directed collaboration at the Académie required yet another shift in the *persona* of the natural philosopher. A balance had to be struck between adherence to the institution and the pursuit of personal interests such as prestige, fame, and property.

The idea of scientific progress helped to reconcile the tension between the collective institution and strong individuals. Reconciliation, at least in its literary form, was apparent in the way Fontenelle conceived the *éloges*, which served as a 'collective biography' of the institution (Grimaldi 2020, 54). Fontenelle presented individual academicians as part of the overall progress, with the help of parallels between the development of the individual and the whole. This is well illustrated in the *éloge* of Cassini, as Mazauric (2007, 281) and Armand (2006, 86) noted.

In his eulogy of Cassini from 1713, Fontenelle mentioned that the famous astronomer had been interested in astrology in his youth. Recounting this episode, Fontenelle did not want to tarnish Cassini's reputation in the perception of his early Enlightenment audience. The point of this mention was to show that Cassini overcame this infantile phase through reason and observation (Fontenelle 1990–2001, vol. 6, 266). Cassini had freed himself from his childish naivety and

[59] Fontenelle 1990–2001, vol. 6, 59; English translation in Fontenelle 1727, xxii.

[60] See Huygens, letter to Doublet, 11 March 1667; Huygens 1888–1950, vol. 6, 119. The letter is quoted by Tits-Dieuaide 1998, 102.

adopted an adult view of the universe. Fontenelle's narrative then continued with how the young Cassini formulated bold hypotheses without having experiences and observations confirming them. Only over time did he realize that he needed many observations and reliable instruments. But then again, he had to face hostility and fight prejudice in the same way as the sciences did after their rebirth at the beginning of the seventeenth century.

The *éloge* of Cassini thus combined ontogeny and phylogeny in a way that framed Cassini's intellectual career as a microhistory of human thought. Similarly to Cassini, humans at the beginning were superstitious, then they invented speculative systems, and only much later they realized that astronomy had to be based on observations and precise instruments. In the course of its history, astronomy had to constantly struggle with prejudices, just like Cassini (Fontenelle and Godin 1733, vol. 1, 1–6). In this way, Cassini's life could be represented as the independent pursuit of individual interests and desires, but at the same time, Fontenelle could show the inevitable connection between individual and collective efforts, since in his view progress had an unchanging structure valid for both, individual and collective development.

The 1699 reform of the Académie purposefully worked with the idea of the institution's longevity and tried to secure the foundations for its *vita longa*. Articles XL and XLI required the secretary to record the proceedings of the academicians and to preserve them in the registers of the Académie. After gathering these records, the secretary was obliged to compose regular reports addressed to the monarch and the public. The manuscript registers of the Académie and the published *mémoires* in the *HARS* volumes had several functions. Understandably, they were to convey the research of the academicians to the king, the court, and the public. Further, Fontenelle created the memory of the institution in the *HARS* volumes; this became the basis of its continuing identity (Mazauric 2007, 83; Seguin 2023).

In addition, registers and *mémoires* fulfilled an epistemological function of bequeathing 'naked' data to future scientists. Daston observed that the purpose of scientific repositories and archives was not to preserve the past but to create a continuity of research. Above all, the data stored were meant to be used by future researchers for processing, comparison, and evaluation resulting in new discoveries. The function of archives was 'to annihilate time', Daston concluded (Daston 2017, 1, 10).

In the case of the Académie, the 'annihilation of time' meant that the existence of registers and *mémoires* created a permanent present. Academicians could look at decades-old observational and experimental results and incorporate them into their research. For example, in his *Histoire des Plantes (1698),* Joseph Pitton de Tournefort referred to chemical analyses that had been carried out for twenty years by Claude Bourdelin. Records of these analyses were stored in registers (Tournefort 1698, unpaginated preface). Through references in the various *HARS* volumes, Fontenelle linked research that was years apart, thus showing the evolution of the problem itself and the continuity of research at the Académie. The historian of science was thus able to participate in the progress of the human spirit by bringing together isolated findings to prepare a future synthesis of all knowledge (Seguin 2023, 13).

Registers, archives and published volumes also embodied the institutional unity of experience with nature by separating it from its individual bearers. At the end of the preface to the *Mémoires pour servir à l'histoire naturelle des animaux*, Claude Perrault articulated the belief in the institutional unification of experience by drawing on the ancient model encountered in Gessner. Alexander had Aristotle, to whom experience flowed from all over the empire; Louis XIV had the Académie, whose collective of personalities would produce a work that would surpass that produced for Alexander (Perrault 1671, fol. ō-v). Perrault thus succeeded in simultaneously highlighting both the importance of the Sun King and denying the ancient model as an epistemological norm. Even if Aristotle's experience was supported by the absolute power of the king, it could not embrace a full variety of nature. Only the institution could collect and preserve a sufficient amount of experience. Fontenelle expressed this view in the *Préface sur l'utilité des mathématiques et da la physique*[61]:

> [N]othing but Societies [...] can succeed in making and preparing it [a collection of data]: Neither the Lights, nor the Cares, nor the Life, nor the Purse of any Private Man are equal to it. There is required too great a Number of Experiments, and of two [*sic*!] many different Kinds, and a too frequent Repetition of the same, too great a Variation of the Manner of 'em, and too much Time to be spent in them for any one Man whatever.

Institutionalization made it possible to homogenize the discontinuity of individual experiences, to unify partial experiences, and to free the process of knowledge from the speculative rush of individuals who believed they could achieve the goals of knowledge in their lifetimes. Although historians have dealt extensively with the institutionalization of natural philosophy in recent years,[62] little mention has been made of the epistemological and anthropological aspects of this change.

From an epistemological point of view, the institution was the means by which finite human beings could master the infinite variety and magnitude of nature. Fontenelle was able to introduce a personified institution as a new bearer of knowledge that possessed a life long enough to know the immeasurable nature: 'The History of Animals, as well as that of Plants, is of almost immense scope & [they] are properly those kinds of works which belong only to academies (*Compagnies*), because they are immortal & they can dispose of as many centuries as they need' (Fontenelle and Godin 1733, vol. 1, 52). With its archives and generational continuity, the institution gave the sciences time that the individual did not have. The Académie thus became a human achievement that was adequately adapted to discovering the full extent of nature.

From an anthropological point of view, the purpose of institutionalization was to make natural philosophy independent of the short-term contingency of individual life and to give it a continuous identity that included historical longevity. Blumenberg noted that

---

[61] Fontenelle 1994, vol. 6, 49; English translation in Fontenelle 1727, xvi.

[62] See e.g. Licoppe 1994, Biagioli 1995, McClellan III 1981.

> the institutionality of science consists precisely in not being affected by the life processes of the subjects who conducted it. This makes the plurality of its bearers indifferent to their theoretical service. The process proceeds as if it were that of a single subject, identical in its motives and goals (Blumenberg 1986, 173).

Scientific institutions could become immortal subjects of knowledge that were not threatened by the self-centredness and short lives of their human crew. Institutions thus created the epistemic environment for the introduction of the idea of scientific progress, which was hardly thinkable without the long-standing subject of history.

I have already suggested that the Académie had to carefully balance the collectivization of research and the satisfaction of individual desires for knowledge and recognition. Thus, the ideal of a long-lived immortal institution was in fact subject to the interests of the mortal members of the Académie. In the late seventeenth and early eighteenth centuries, there were several family clans at the Académie; Sturdy described them in detail (Sturdy 1995, 193–212, 243–250, 321–324). The most famous was the Cassini family. In 1694, the 17-year-old Jacques Cassini was elected to the Académie. The first son of the astronomer and mathematician La Hire entered the institution in the same year, and the second son became an academician in 1710. Other academic families were the Maraldis, Couplets, Lemérys, and Geoffroys.

As Sturdy has shown, there is no doubt that this practice was pure nepotism, serving the benefit of individual families. Academicians protected their sons (or nephews, or brothers) and took up institutional positions for them that could have been filled by other, perhaps more gifted, persons (Sturdy 1995, 413). On the other hand, president Bignon, secretary Fontenelle, and minister Pontchartrain must have known about this practice, so it was probably a purposeful and deliberate strategy.

Before the 1699 reform, scholars entered the academy generally as men of a mature age who had already proven themselves to be eminent scientists and philosophers. However, relatives of the academicians entered the institution as boys or young men. This succession arguably reflected a change in the logic of the institution, which began to purposefully educate its own members and shape their epistemological settings from a young age. Sturdy speculated that the Académie had a great deal of specific equipment and expensive apparatus at its disposal, among which the offspring of academicians moved from a young age. This was especially true of the Observatoire, where the Cassinis, Maraldis, and La Hires lived. The academicians and the administrators of the Académie might have understood the presence of the male teenagers at the Académie's sites as an opportunity to ensure the continuity and coherence of research at the Académie (Sturdy 1995, 242). They seem to consider such a succession the most obvious option. In the *HARS* volume for the academic year 1702, Fontenelle announced that the position of the deceased Nicolas Lemery was naturally (*naturellement*) taken over by his son Louis (Fontenelle 1702–1742, vol. 4, 139).

In a cultural context in which there was limited education in modern natural philosophy, the Académie apparently did not want to wait for mature and distinguished scientists but began to educate young men to ensure its future flourishing: The boys could be inculcated from the beginning with appropriate epistemological standards and epistemic practices. The fact that they were members of academic families suggests that the Académie at this time lacked a model of intergenerational exchange that would ensure continuity of research, and therefore resorted to a traditional form of family succession. To some extent, the academicians still imagined institutional immortality as genetic immortality.

Bibliography

Armand, Guilhem. 2006. La construction de l'image du savant chez Fontenelle. *Revue Fontenelle* 4: 79–93.

Armogathe, Jean-Robert. 2001. Une Ancienne Querelle. In *La Querelle des Anciens et des Modernes. XVIIe-XVIIIe siècles*, ed. Anne-Marie Lecoq, 801–849. Paris: Gallimard.

Ashworth, William B. 1990. Natural history and the emblematic world view. In *Reappraisals of the scientific revolution*, ed. David Lindberg and Robert S. Westman, 303–332. Cambridge.

Audidière, Sophie. 2016. Introduction: Fontenelle ou la tendresse philosophe. In Bernard de Fontenelle, *Digression sur les Anciens et les Modernes et autres textes philosophiques*, ed. Sophie Audidière, 13–58. Paris: Garnier.

Bacon, Francis. 1857–1874. In *The works of Francis Bacon*, ed. James Spedding, 14 vols. London: Longman.

———. 1996–2008. *The Oxford Francis Bacon*, ed. Graham Rees et al., 15 vols. Oxford: Clarendon Press.

Baillet, Adrien. 1691. *La vie de Monsieur Descartes*, 2 vols. Paris: Horthemels.

Belon, Pierre. 1551. *L'histoire génerale des estranges poissons marins*. Paris: Chaudiere.

———. 1555. *L'histoire de la nature des oyseaux*. Paris: Benoist Preuost.

Biagioli, Mario. 1995. Le prince et les savants. La civilité scientifique au 17e siècle. *Annales* 50: 1417–1453.

Blair, Ann. 1992. Humanist methods in natural philosophy. *Journal of the History of Ideas* 53: 541–551.

———. 2003. Reading strategies for coping with information overload, ca. 1450–1700. *Journal of the History of Ideas* 64: 11–28.

Blechet, Françoise. 1989. Fontenelle et l'abbé Bignon: Du Président de l'Académie Royale des Sciences au Secrétaire Perpetuel. *Corpus* 13: 51–62.

Blumenberg, Hans. 1965. *Kopernikus im Selbstverständnis der Neuzeit*. Wiesbaden: Harrassowitz.

———. 1986. *Lebenszeit und Weltzeit*. Frankfurt am Main: Suhrkamp.

———. 1996. *Die Legitimität der Neuzeit. Erneuerte Ausgabe*. Frankfurt am Main: Suhrkamp.

Bodin, Jean. 1597. *Universae naturae theatrum*. Francofurti: Wechelius.

Bossuet, Jacques-Bénigne. 1881. *De la connaissance de Dieu et de soi-même*, ed. Charles Jourdain. Paris: Firmin-Didot.

Boyle, Robert. 1775. *The works*. ed. Thomas Birch. 6 vols. London: Rivington.

Brahe, Tycho. 1913–1929. *Tychonis Brahe Dani opera omnia*, ed. John L. Dreyer, 10 vols. Copenhagen: Gyldendal.

———. 1996. *Instruments of the renewed astronomy*. translated by H. Raeder, Alena Hadravová, Petr Hadrava, Jole Shackelford. Prague: Koniasch Latin Press.

Bruno, Giordano. 1977. *The Ash Wednesday supper*. translated by Edward A. Gosselin and Lawrence S. Lerner. Toronto: Univeristy of Toronto Press.

Bury, John B. 1920. *The idea of progress. An inquiry into its origin and growth*. London: Macmillan.

Campanella, Tommaso. 1622. *Apologia pro Galileo*. Francofurti: Kempferus.

Cardano, Girolamo. 1663. *Opera omnia*. 10 vols. Lugduni: Huguetan.

Cassini, Jean-Dominique. 1673. *Découverte de deux nouvelles planetes autour de Saturne*. Paris: Mabre-Cramoisy.

———. 1681. *Observations sur la comete qui a paru au mois de decembre 1680*. Paris: Michallet.

———. 1684. *Les Élémens de l'astronomie vérifiez par M. Cassini*. Paris: L'imprimerie royale.

Castel de Saint-Pierre, Charles I. 1737. *Observasions sur le progrèz continuel de la Raizon Universelle*. In Charles I. Castel de Saint Pierre, *Ouvrajes de politique*, vol. 11. Paris: Rotterdam.

Čermáková, Lucie, and Jana Černá. 2018. Naked in the old and the new world: Differences and analogies in descriptions of European and American herbae nudae in the sixteenth century. *Journal of the History of Biology* 51: 69–106.

Christianson, John R. 2000. *On Tycho's Island: Tycho Brahe and his assistants, 1570–1601*. Cambridge: Cambridge University Press.

Clarke, Desmond. 1982. *Descartes' philosophy of science*. Manchester: Pennsylvania State University Press.

———. 1989. *Occult powers and hypotheses. Cartesian natural philosophy under Louis XIV*. Oxford: Oxford University Press.

———. 1992. Descartes' philosophy of science and the scientific revolution. In *The Cambridge companion to Descartes*, ed. John Cottingham, 259–285. Cambridge: Cambridge University Press.

Comenius, Johannes A. 1651. *A paterne of universal knowledge*. translated by Jeremy Collier. London: Collins.

———. 1964. *De rerum humanarum emendatione consultatio catholica*, 2 vols. Academia: Pragae.

Copenhaver, Brian. 1978. The historiography of discovery in the Renaissance: The sources and compositions of Polydore Vergil's *De inventoribus rerum*, I-III. *Journal of the Warburg and Courtald Institutes* 41: 192–214.

Copernicus, Nicolaus. 1975. *De revolutionibus*. In Nicolaus Copernicus, *Opera omnia*, ed. Ryszard Gansiniec, vol. 2. Warsaw and Cracow: Officina publica libris scientificis edendis.

———. 1978. On the revolutions. In *Nicolaus Copernicus, complete works*, ed. Jerzy Dobrzycki translated by Edward Rosen. Warsaw/Cracow: Polish Scientific Publishers.

Cottingham, John. 1992. A new start? Cartesian metaphysics and the emergence of modern philosophy. In *The rise of modern philosophy*, ed. Tom Sorell, 145–166. Oxford: Clarendon Press.

Daston, Lorraine. 1995. The moral economy of science. *Osiris* 10: 3–24.

———. 2011. The empire of observation, 1600–1800. In *Histories of scientific observation*, ed. Lorraine Daston and Elisabeth Lunbeck, 81–135. Chicago/London: The University of Chicago Press.

———. 2017. Introduction: Third nature. In *Science in the archives. Pasts, presents, futures*, ed. Lorraine Daston, 1–16. Chicago/London: The University of Chicago Press.

De Smet, Rudolf, and Karin Verelst. 2001. Newton's scholium generale: The Platonic and Stoic legacy—Philo, Justus Lipsius and the Cambridge Platonists'. *History of Science* 39: 1–30.

Dear, Peter. 1985. Totius in verba: Rhetoric and authority in the early Royal Society. *Isis* 76: 146–157.

———. 2015. Cultures expérimentales. In *Histoire des sciences et des savoirs. Vol. 1: De la Renaissance aux Lumières*, ed. Stéphane Van Damme. Paris: Seuil.

Delvaille, Jules. 1910. *Essai sur l'histoire de l'idée de progrès. Jusqu'a la fin du XVIIIe siècle*. Paris: Félix Alcan.

Descartes, René. 1984–1991. *The philosophical writings*, 3 vols. translated by John Cottingham, Robert Stoothoff, Dugald Murdoch, and Athony Kenny Cambridge: Cambridge University Press.

———. 1996. In *Œuvres de Descartes*, ed. Charles Adam and Paul Tannery, 11 vols. Paris: Vrin.

Dew, Nicolas. 2015. Un Colbertisme scientifique? In *Histoire des sciences et des savoirs. 1. De la renaissance aux Lumières*, ed. Stéphane Van Damme, 431–445. Paris: Seuil.

Dodart, Denis. 1679. *Mémoires pour servir a l'histoire des plantes*. Paris: L'Imprimerie royale.

Dodoens, Rembert. 1577. Studiosis Medicinae Candidatos. In Rembert Dodoens, *Histoire des plantes*. Non-paginated preface. Anvers: Loë.

Dreyer, John L.E. 1890. *Tycho Brahe. A picture of scientific life and work in the sixteenth century*. Edinburgh: Black.

Dunphy, Graeme. 2010. Six ages of the world. In *Encyclopedia of the medieval chronicle*, ed. Graeme Dunphy, 1367–1370. Leiden: Brill.

Duris, Pascal. 2016. *Quelle Révolution Scientifique? Les sciences de la vie dans la querelle des Anciens et des Modernes (XVIe-XVIIIe siècles)*. Paris: Hermann.

Findlen, Paula. 2006. Natural history. In *The Cambridge history of science. Vol. 3: Early modern science*, ed. Lorraine Daston and Katherine Park, 435–468. Cambridge: Cambridge University Press.

Fontenelle, Bernard de 1702–1742. *Histoire de l'Académie royale des sciences*, 40 vols. Paris: Boudout.

———. 1727. *Lives of the French, Italian and German philosophers, late members of the royal academy of sciences in Paris*. London: Innys.

———. 1960. A digression on the ancients and the moderns (1688), translated by Donald Schier. In *The continental model. Selected French critical essays of the seventeenth century in English translation*, ed. Scott Elledge and Donald Schier, 358–372. Minneapolis: Carleton College.

———. 1990. *Conversations on the plurality of worlds*, translated by H. A. Hargreaves. Berkeley: University of California Press.

———. 1990–2001. In Bernard de Fontenelle, *Œuvres complètes*, ed. Alain Niderst, 9 vols. Paris: Fayard.

———. 1994. Préface sur l'utilité des mathématiques et de la physique. In Bernard de Fontenelle, *Œuvres complètes*, ed. Alain Niderst, vol. 6. Paris: Fayard.

Fontenelle, Bernard de, and Louis Godin. 1733. *Histoire de l'Académie Royale des sciences, depuis son Etablissement en 1666 jusqu'à 1699*, 2 vols. Paris: Martin, Coignard & Guerin.

Friedrich, Udo. 1995. *Naturgeschichte zwischen artes liberales und frühneuzeitlicher Wissenschaft. Conrad Gessners 'Historia animalium' und ihre volkssprachige Rezeption*. Tübingen: Niemeyer.

Frisch, Andrea. 2014. Passing knowledge: André Thévet's cosmographical epistemology. *Journal of Early Modern History* 18: 49–67.

Galilei, Galileo. 1890–1909. *Edizione nazionale delle opere di Galileo Galilei*, ed. Antonio Favaro, 20 vols. Firenze: Barbèra Editore.

Galilei, Galileo, and Christoph Scheiner. 2010. *On sunspots*, translated by Eileen Reeves and Albert van Helden. Chicago and London: University of Chicago Press.

Gassendi, Pierre. 1658. *Opera omnia*, 6 vols. Lugduni: Anisson.

Gatz, Bodo. 1967. *Weltalter, goldene Zeit und sinnverwandte Vorstellungen*. Hildesheim: Olms.

Gaukroger, Stephen. 2006. The persona of natural philosopher. In *The philosopher in early modern Europe. The nature of a contested identity*, ed. Conal Condren, Stephen Gaukroger, and Ian Hunter, 17–34. Cambridge: Cambridge University Press.

Gembicki, Dieter, and Rolf Reichardt. 1993. Progrès. In *Handbuch politisch-sozialer Grundbegriffe in Frankreich, 1680–1820. Heft 14–15: Corruption. Décadence. Ordre. Désordre. Progrès*, 101–153. Berlin/Boston: Oldenbourg.

Gessner, Conrad. 1551–1555. *Historia animalium*, 3 vols. Tiguri: Froschoverus.

———. 1552. Praefatio. In Hieronymus Tragus, *De stirpium differentiis*. Strassburg: Wendel Rihel.

———. 1577. *Epistolarum medicinalium libri III*. Tiguri: Froschoverus.

Gingerich, Owen, and James R. Voelkel. 1998. Tycho Brahe's Copernican campaign. *Journal for the History of Astronomy* 29: 1–34.

Glanvill, Joseph. 1668. *Plus ultra, or the progress and advancement of knowledge since the days of Aristotle*. London: Collins.

———. 1885. In *Scepsis scientifica, or vanity of dogmatizing*, ed. James Owen. London: Kegan Paul, Trench & Company.

Graf-Stulhofer, Franz. 2012. Tradition(en) und Empirie in dr frühneuzeitlichen Naturforschung. In *Wissenschaft und Kultur an der Zeitwende*, ed. Helmuth Grössing and Kurt, 63–80. Göttingen: Mühlberger.

Grafton, Anthony. 1992. Kepler as a reader. *Journal of the History of Ideas* 53: 561–572.

Grimaldi, Claudio. 2020. *Les Éloges de Fontenelle*. Paris: L'Harmattan.

Guyer, Ferdinand E. 1921. C'est nous qui sommes les anciens. *Modern Language Notes* 36: 257–264.

Hadot, Pierre. 2006. *The veil of Isis. Essay on the history of the idea of nature*. Cambridge, MA/London: Harvard University Press.

Hahn, Roger. 1971. *The anatomy of a scientific institution. The Paris Academy of Sciences, 1666–1803*. Berkeley/Los Angeles: University of California Press.

Hakewill, George. 1627. *An apologie of the power and providence of God*. Oxford: Lichfield/Turner.

Harrison, Peter. 2006. The natural philosopher and the virtues. In *The philosopher in early modern Europe. The nature of a contested identity*, ed. Conal Condren, Stephen Gaukroger, and Ian Hunter, 202–228. Cambridge: Cambridge University Press.

———. 2011. Natural history. In *Wrestling with nature. From omens to science*, ed. Peter Harrison, Ronald L. Numbers, and Michael H. Shank, 117–148. Chicago/London: University of Chicago Press.

Hevelius, Johannes. 1673. *Machina coelestis*, 2 vols. Gedani: Reiniger.

Hippocrates. 1844. *Aphorismes*. In *Oeuvres complètes de Hippocrate*, I-X, ed. E. Littré. Paris: Hakkert, vol. 4.

Hobbes, Thomas. 1996. *Leviathan*, ed. J.C.A. Gaskin. Oxford/New York: Oxford University Books.

Hooke, Robert. 1705. *Posthumous works*. London: Waller.

Huygens, Christiaan. 1888–1950. *Œuvres complètes*, 22 vols. La Haye: Nijhoff.

Ianziti, Gary. 2012. *Writing history in renaissance Italy: Leonardo Bruni and the uses of the past*. Cambridge, MA/London: Harvard University Press.

Kamlah, Wilhelm. 1961. Der Anfang der Vernunft bei Descartes—autobiographisch und historisch. *Archiv für Geschichte der Philosophie* 43: 70–85.

Kepler, Johannes. 1938–2009. *Gesammelte Werke*, ed. M. Caspar, V. Bialas et al., 25 vols. München: C. H. Beck.

———. 1972. Preface to the *Rudolphine tables*. translated by Owen Gingerich and William Walderman. *Quarterly Journal of the Royal Astronomical Society* 13: 360–373.

Koselleck, Reinhart. 2006. *Begriffsgeschichten. Studien zur Semantik und Pragmatik der politischen und sozialen Sprache*. Frankfurt am Main: Suhrkamp.

Leinkauf, Thomas. 2017. *Grundriss: Philosophie des Humanismus und der Renaissance*, 2 vols. Hamburg: Felix Meiner.

Levitin, Dmitri. 2015. *Ancient wisdom in the age of the new science. Histories of philosophy in England, c. 1640–1700*. Cambridge: Cambridge University Press.

Leyden, W. Von. 1958. Antiquity and authority: A paradox in the Renaissance theory of history. *Journal of the History of Ideas* 19: 473–792.

Licoppe, Christoph. 1994. The crystallization of a new narrative form in experimental reports (1660–1690). The experimental evidence as a transaction between philosophical knowledge and aristocratic power. *Science in Context* 7: 205–224.

Malebranche, Nicolas. 1979. *De la recherche de la vérité*. In Malebranche, Nicolas. *Oeuvres*, ed. Geneviève Rodis-Lewis, vol. 1. Paris: Gallimard.

Mazauric, Simone. 1998. *Gassendi, Pascal et la querelle du vide*. Paris: Presses Universitaires de France.

———. 2007. *Fontenelle et l'invention de l'histoire des sciences à l'aube des Lumières*. Paris: Fayard.

McClellan III, James. 1981. The Académie Royale des Sciences, 1699–1793: A statistical portrait. *Isis* 72: 541–567.

———. 2003. Scientific institutions and the organization of science. In *The Cambridge history of science. Vol. 4: Eighteenth-century science*, ed. John Porter, 87–106. Cambridge: Cambridge University Press.

Mémoires de l'Académie royale des sciences, depuis 1666 jusqu'à 1699. 1729–1734. 11. vols. Paris: Compagnie des Libraires.

Mémoires de mathématique et de physique, tirez des registres de l'Académie royale des sciences. 1702–1742, 40 vols. Paris: Boudot (the same volumes as Fontenelle 1702–1742 but with separate pagination).

Mosley, Adam. 2007. *Bearing the heavens: Tycho Brahe and the astronomical community of the late sixteenth century*. Cambridge: Cambridge University Press.

Motto, Anna L. 1984. The idea of progress in Senecan thought. *The Classical Journal* 79: 225–240.

Müller, Jan-Dirk. 1986. Erfahrung zwischen Heilssorge, Selbsterkenntnis und Entdeckung des Kosmos. In *Literatur und Kosmos*, ed. Williams G. Scholz, 307–342. Amsterdam: Rodopi.

———. 1991. Alt und neu in der Epochenerfahrung um 1500. Ansätze zur kulturgeschichtlichen Periodisierung in frühneuhochdeutschen Texten. In *Traditionswandel und Traditionsverhalten*, ed. Walter Haug and Burghart Wachinger, 121–144. Berlin/New York: Niemeyer.

———. 1998. Universalbibliothek und Gedächtnis. Aporien frühneuzeitlichen Wissenskodifikation bei Conrad Gessner. In *Erkennen und Erinnern in Kunst und Literatur*, ed. Wolfgang Frühwald, Dietmar Peil, Michael Schilling, and Peter Strohschneider, 285–310. Berlin/Boston: De Gruyter.

Neuser, Wolfgang. 2011. Fortschritt. In *Neues Handbuch der philosophischer Begriffe*, ed. Petra Kolmer and Armin G. Wildfeuer, vol. 1, 797–801. München: Karl Albert.

Niderst, Alain. 1972. *Fontenelle à la recherche de lui-même (1657–1702)*. Paris: Nizet.

Ogilvie, Brian W. 2003. The many books of nature: Renaissance naturalists and information overload. *Journal of the History of Ideas* 64: 29–40.

———. 2006. *The science of describing. Natural history in Renaissance Europe*. Chicago/London: The University of Chicago Press.

Olmsted, John W. 1942. The scientific expedition of Jean Richer to Cayenne (1672–1673). *Isis* 34: 117–128.

Oviedo, Fernando G. 1555. *L'histoire naturelle et generalle des Indes*. Paris: Vascosan.

Panciroli, Guido. 1655. *Libri duo rerum jam olim deperditarum et noviter inventarum*. Francofurti: Schwonweterius.

Pascal, Blaise. 1910. Preface to the treatise on vacuum. translated by O. W. Wight. In Blaise Pascal, *Thoughts. Letters. Minor works*, 440–450. New York: Collier.

———. 1998. Préface sur le Traité du vide. In Blaise Pascal, *Œuvres complètes*, ed. Michel Le Guern, vol. 1. Paris: Gallimard.

Perrault, Claude. 1671. *Mémoires pour servir à l'histoire naturelle des animaux*. Paris: L'imprimerie royale.

Perrault, Charles. 1688–1697. *Parallèle des anciens et des modernes*, 4 vols. Paris: Coignard.

Picard, Jean, and Philippe de La Hire. 1729. Observations Astronomiques faites en divers endroits du royaume par Monsieur Picard. In *Mémoires de l'Academie Royale des sciences*, vol. 7, 358–429. Paris: Compagnie des Libraires.

Pinon, Laurent. 1995. *Livres de Zoologie de la Renaissance. Une anthologie (1450–1700)*. Paris: Klincksieck.

———. 2003. Entre compilation et observation: l'ecriture et l'ornithologie d'Ulisse Aldrovandi. *Genesis* 20: 53–70.

———. 2005. Conrad Gessner and the historical depth of Renaissance natural history. In *Historia. Empiricism and erudition in early modern Europe*, ed. Gianna Pomata and Nancy G. Siraisi, 241–267. Cambridge, MA/London: The MIT Press.

Pliny. 1967. *Natural history*. translated by Harris Rackham. Vol. 3, libri VIII-XI. Cambridge, MA/London: Harvard University Press/Heinemann.

Poulouin, Claudine. 1998. *Les temps des origines. L'Eden, le Déluge et 'les temps reculés'. De Pascal à l'Encyclopédie*. Paris: Honoré Champion.

———. 2006. La Préface sur l'utilité des mathématiques et de la physique de Fontenelle ne vise-t-elle qu'à rendre le progrès des sciences visible au 'gros du monde'? *Revue Fontenelle* 4: 61–78.
Power, Henry. 1664. *Experimental philosophy*. London: Roycrofe.
Ptolemy, Claudius. 1998. *Ptolemy's Almagest*. Translated and ed. Gerald J. Toomer. Princeton: Princeton University Press.
Ray, John. 1735. *The wisdom of god*. London: Innys/Manby.
Rheticus, Joachim. 1971. *Narratio prima*. In *Three Copernican Treatises*, ed. Edward Rosen. New York: Octagon.
———. 1982. In Joachim Rheticus, *Narratio prima*, ed. Henri Hugonnard-Roche and Jean-Pierre Verdet. Wroclaw: Ossolineum.
Riedl-Dorn, Christa. 1989. *Wissenschaft und Fabelwesen. Ein kritischer Versuch über Conrad Gessner und Ulisse Aldrovandi*. Wien/Köln: Böhlau.
Rohault, Jacques. 1671. *Traité de physique*. Paris: Thierry.
Rohault, Jacques, and Samuel Clarke. 1735. *System of natural philosophy illustrated with Dr. Samuel Clarke's notes taken mostly out of Sir Isaac Newton's philosophy*, 2 vols. London: John – Knapton.
Rohbeck, Johannes. 1987. *Die Fortschrittsidee der Aufklärung*. Frankfurt am Main/New York: Campus.
Rondelet, Guilaume. 1558. *L'Histoire entiere des poissons*. Lion: Bonhome.
Rose, Paul L. 1975. *The Italian renaissance of mathematics. Studies on humanists and mathematicians from Petrarch to Galilei*. Genève: Librairie Droz.
Rossi, Paolo. 1970. The idea of scientific progress. In Paolo Rossi, *Philosophy, technology and the arts in the early modern era*, 63–99. London: Harper.
Rouvillois, Frédéric. 2010. *L'invention du progrès, 1680–1730*. Paris: CNRS.
Ruellet, Aurélien. 2016. *La Maison de Salomon: Histoire du patronage scientifique et technique en France et Angleterre au XVIIe siècle*. Rennes: Presses Universitaires de Rennes.
Schlobach, Jochen. 1980. *Zyklentheorie und Epochenmetaphorik. Studien zur bildlichen Sprache der Geschichtsreflexion in Frankreich von der Renaissance bis zur Frühaufklärung*. München: Fink.
———. 2001. Progress. In *Encyclopedia of the Enlightenment*, ed. Michel Delon, 1100–1105. London/New York: Routledge.
Schmidt, Roderich. 1955–1956. Aetates mundi. Die Weltalter als Gliederungsprinzip der Geschichte. *Zeitschrift für Kirchengeschichte* 67: 288–315.
Schouls, Peter. 1987. Descartes and the idea of progress. *History of Philosophy Quarterly* 4: 423–433.
Seguin, Maria-Susana. 2022. Anciens et Modernes à l'Académie des sciences. In *Anciens et Modernes face aux pouvoirs. L'Église, le Roi, les Académies (1687–1750)*, ed. Christelle Bahier-Porte and Delphine Reguig, 179–195. Paris: Honoré Champion.
———. 2023. The royal academy of sciences of Paris and the birth of historical epistemology. In *Ordering knowledge: Disciplinarity and the shaping of European modernity*, ed. Jean-Jacques Chardin, 187–208. Strasbourg: Presses universitaires de Strasbourg.
Seneca, Lucius A. 2014. *Natural questions*. Translated Harry M. Hine. Chicago/London: The University of Chicago Press.
Shank, John B. 2018. *Before Voltaire: The French origins of 'Newtonian' mechanics, 1680–1715*. Chicago/London: The University of Chicago Press.
Shapin, Steven. 2000. Descartes the doctor: Rationalism and its therapies. *The British Journal for the History of Science* 33: 131–154.
Špelda, Daniel. 2013. The search for antediluvian astronomy: Sixteenth- and seventeenth-century astronomers' conceptions of the origin of science. *Journal for the History of Astronomy* 44: 337–362.
———. 2022. Collective empiricism at the Paris Observatoire in the late seventeenth century. *The Seventeenth Century* 37: 977–1007.
Sprat, Thomas. 1734. *The history of the royal society of London*. London: Knapton.

Stroup, Alice. 1990. *A company of scientists: Botany, patronage, and community at the seventeenth-century Parisian Royal académie of sciences*. Berkeley/Los Angeles: University of California Press.

Sturdy, David. 1995. *Science and social status: The members of the Académie des sciences (1666–1750)*. Woodbridge: The Boydell Press.

Taguieff, Pierre. 2002. L'idée de progrès: Une approche historique et philosophique. *Les Cahiers du CEVIPOF* 32: 1–136.

Thévet, André. 1558. *Les singularitéz de la France Antarctique, avtrement nommée Amerique*. Anvers: Plantin.

Thoren, Victor. 1990. *The Lord of Uraniborg: A biography of Tycho Brahe*. Cambridge: Cambridge University Press.

Tits-Dieuaide, Marie-Jeanne. 1998. Les savants, la société et l'État: à propos du renouvellement de l'Académie royale des sciences (1699). *Journal des savants* 1: 79–114.

Tournefort, Joseph P. 1698. *Histoire de plantes quie naissent aux environs de Paris*. Paris: L'Imprimerie Royale.

Van Helden, Albert. 1989. The telescope and cosmic dimensions. In *The general history of astronomy, Vol. 2: Planetary astronomy from the Renaissance to the rise of astrophysics. Part A: Tycho Brahe to Newton*, ed. René Taton and Curtis Wilson, 106–118. Cambridge: Cambridge University Press.

———. 1994. Telescopes and authority from Galileo to Cassini. *Osiris* 9: 8–29.

Vergil, Polydore. 2002. In *De inventoribus rerum/On discovery*, ed. Brian Copenhaver. Cambridge, MA: Harvard University Press.

Werle, Dirk. 2007. *Copia librorum. Problemgeschichte imaginierter Bibliotheken, 1580–1630*. Tübingen: Niemayer.

Wilkins, John. 1802. *The mathematical and philosophical works*. London: Vernon and Hood.

Williams, Gareth D. 2012. *The cosmic viewpoint. A study of Seneca's natural questions*. Oxford/New York: Oxford University Press.

Yeo, Richard. 2018. Hippocrates' complaint and the scientific ethos in early modern England. *Annals of Science* 75: 73–96.

Zaho, Ann M. 2004. *Imago Triumphalis. The function and significance of triumphal imagery for Italian renaissance rulers*. New York: Lang.

Chapter 6
Progress of the Human Spirit

Contents

Abstract In the Renaissance context, a cyclical conception of the history of knowledge prevailed, according to which knowledge in the past had undergone several phases of growth and decline. At the end of the seventeenth century, the secretary of the Académie des sciences, Bernard de Fontenelle, introduced a new conception of the history of natural knowledge. He replaced the cyclical conception with a linear and cumulative one. Combining epistemology with the philosophy of history, Fontenelle argued that the historical development of human knowledge correlated with stages in the epistemological advancement of the individual: The epistemological phylogeny repeated the cognitive ontogeny. Fontenelle referred to the epistemological development of humankind as the progress of the human spirit and linked it to social history. Early societies had developed their knowledge only as fables based on imagination, anthropomorphism, and superstition. In the Middle Ages, humans were subject to blind faith in authority. In his view, it was only in modern times that natural philosophers had emancipated themselves from prejudice and begun to develop natural philosophy based on reason and experience, searching for general principles and laws. For Fontenelle, the progress of the human spirit was methodical, cumulative, linear, necessary, and lawful, and it avoided contingent skips in its growth. This conception also redefined the assessment of ancient natural philosophy. Bacon and Descartes had strictly rejected Greek philosophy; in Fontenelle's conception, ancient natural philosophy was the first and initial stage in the long-term process that natural philosophy had to go through. Fontenelle's colleagues from the Académie des sciences were rather reluctant to articulate their

D. Špelda, *The Origins of the Idea of Scientific Progress*, International Archives of the History of Ideas Archives internationales d'histoire des idées 250,
https://doi.org/10.1007/978-3-031-60526-0_6

views on the history of the sciences; therefore, Fontenelle considered his task precisely to embed the specific reports of academicians in his overall narrative of the history of the human spirit. All the specific scientific achievements became parts of a large historical narrative.

Keywords Bernard de Fontenelle · Cyclical philosophy of history · Linear philosophy of history · Idea of progress · Académie des sciences · Quarrel of the ancients and the moderns

The subject for the working of the temporal regime of knowledge was the institution that made it possible to unify and preserve the experience of individuals. However, one more condition had to be met for the emergence of the idea of scientific progress. The time of scientific research had to be connected with human time. The epistemic practice of the temporal regime of knowledge had to be inscribed in history in order to determine a beginning, a course, and a function to the advancement of sciences.

Integrating epistemology with history was important for two reasons. Firstly, early modern culture often understood itself and its history through models from classical Greek and Roman literature. Therefere, the sciences had to be related specifically to classical antiquity by a convincing historical account that gave credit to both the Moderns and the Ancients. Secondly, a historical narrative had to be constructed that substantiated the importance and function of the sciences for society showing the emancipatory force of knowledge.

In this chapter I want to show that Fontenelle was able to transfer the temporal regime of knowledge to history through a deliberate historicization of epistemological processes. He thus produced a grand narrative of the progress of the human spirit that allowed the existing sciences to be interpreted as the inevitable outcome of history while opening up the perspective of further advancement. At the end of the seventeenth century, such a philosophy of history was neither self-evident nor universally accepted. Theologians, libertines, and humanistically educated scholars such as poets or writers held a different view of history. To understand the inherent tension between these different forms of interpreting the history of learning, it is necessary to begin this chapter with the Renaissance conception of the history of knowledge.

6.1 Cyclical History of Knowledge

Although in late antiquity, Augustinus and other Church fathers introduced the Christian view of history, this did not mean the demise of the cyclical conception of history that abounded in Greek philosophy and literature (Kehl and Marrou 1978; Demandt 2011). Krzysztof Pomian asserted that in the Middle Ages, linear and irreversible time only applied to the sacred history and to the history of the Church, which was a supernatural institution that resisted the destructive effects of time. In contrast, profane history was seen as a succession of phases of flourishing and decline that affected empires and states in particular. Pomian also explained that the Church was unique in its linear progression, whereas cyclical profane history was plural. Several historical cycles of profane processes were occurring concurrently and were in different stages of development (Pomian 1988, 39–40).

During the Renaissance, this division applied not only to the history of empires. While the linear view remained valid in eschatology, the cyclical model was naturally transferred to knowledge due to the discovery of the medieval rupture between classical antiquity and the period that started to understand itself as epoch of 'rebirth'. Many Renaissance scholars were convinced that God revealed the truth to humans at the beginning of history and then a decline or cycles of decline and resurrection followed (Findlen 2002). The cyclical concept naturally coincided with the self-understanding of the Renaissance as a period of great *renovatio*, coming after a long period of barbarism. The abundant use of the terms *restitutio, instauratio, restauratio, emendatio* in Renaissance writing originated in conviction that historical change was cyclical, not linear. The change was the return of the old, not the arrival of the brand new (Burke 1980; Schlobach 1980, 97–192). The Renaissance scholars thus often saw their period as the beginning of another cycle which brought back culture that had been lost.

Following this cyclical paradigm, the Renaissance Platonists (and other authors as well) presented their philosophy as a return to the original knowledge of the beginning of history, held by the sages and biblical figures to whom exceptional knowledge was attributed. The knowledge of ancient times, *prisca theologia* or *prisca sapientia*,[1] was considered closed, finished, complete, perfect, and truthful. The truth was revealed to humans at the beginning of history in its completeness and was not further changed. It had only been more or less successfully passed down through history (Ficino 2001–2006, vol. 1, 116–118). *Prisca sapientia* was considered a closed set of teachings that the inspired sage passed on to the disciples to be further propagated. The disciples could further dilute the truth handed down, but they could not add anything substantial to it because human reason could not surpass the truth revealed by God.

For the Renaissance Platonists, the truth was everlasting, but the number of people who were enlightened by it was changing. The ancient truth was repeatedly forgotten and renewed throughout history, and it was generally believed that this process was directed by Divine Providence according to a plan unknown to humans. In Ficino's view, history alternated between enlightened periods and 'unhappy ages' (*saecula infelicia*), when true knowledge was forgotten and covered over by barbaric ignorance and superstition.[2] The advent of these periods was neither accidental nor necessary. Providence was always watching over the process of transmitting the truth, and it never allowed the truth to disappear completely.[3] The idea of the gradual

[1] Vasoli 2002; cf. Walker 1972, 20–21; Idel 2002; Hanegraaff 2005; Tambrun 2006.

[2] Ficino 1576, 854, 925; cf. Hankins 1991, vol. 1, 181.

[3] Ficino 1576, 871; cf. 1128, 1130, 1537.

accumulation of true knowledge through autonomous human activity over time did not appear in this type of historical mentality.

Platonists like Ficino were usually convinced that they were in an ascending phase, when, through the rediscovered *prisca sapientia*, classical learning would be revived and saved from harmful influences. Ficino's conception of the history of knowledge was thus approximate to a sinusoid. Periods of flourishing, characterized by a harmonious blend of philosophy and religion, alternated with periods of decline, when foolish superstition and philosophical impiety reigned supreme (Ficino 1576, 871, cf. 1128–1130). In this conception, knowledge did not need time, because it was available from the beginning. It just emerged and disappeared in cycles of flourishing, wickedness, and barbarism. Similarly, other Renaissance writers saw history as having phases of flourishing and decline, which modern interpreters likened to sinusoids or oscillations (Pomian 1988, 47, 53; Schlobach 1980, 164–167).

Following the common understanding of his time, Giordano Bruno used the term *vicissitudo* to describe the constant changeability of the created world. It was precisely in its mutability that the world differed from the intellectual realm of truth, which was eternal and constant. Paul R. Blum explained that because of the immutability of truth, in Bruno's thought, there was no room for the idea of progress. Throughout history, there were periods of flourishing and decline during which truth emerged and disappeared again (Blum 1999). In Bruno's view, the truth of the heliocentric order of the heavens had been known to the Pythagoreans; Copernicus only revived it (Bruno 1977, 29). Blum further pointed out that the reappearances of truth were related to the idea of the eternal recurrence of the same, which Bruno liked to articulate using the biblical phrase *nihil sub sole novum* (Blum 2012, 80–90; cf. McKnight 1991, 114–120). Thus, there was no cumulative growth of the truth in history, only its reappearance. This also led Bruno to a marked distinction between old and new knowledge: 'there is nothing which may not yet be old, and nothing old which has not once been new' (Bruno 1977, 43). In such a conception, it was hardly possible to think of the progress of knowledge.

Not all Renaissance authors understood the history of knowledge as a return of the same immutable truth. In the eleventh book of his treatise *De la Vicissitude ou Variété des choses en l'universe* (1575), Loys Le Roy extolled the discoveries of his time and highlighted the Renaissance triptych of inventions (the compass, the printing press, and gunpowder) as examples of its superiority (Le Roy 1575, 99v–100v). Le Roy also used Senecan metaphors and analogies to formulate his ideas about the progress of knowledge: the beginnings of the sciences are the most difficult, great things come late and with difficulty, posterity will know much that we do not yet know (Le Roy 1575, 115v).

Because of these passages, some historians consider Le Roy's works to represent the modern idea of progress.[4] In fact, Le Roy worked with the traditional cyclical view of history which he combined with the descriptive idea of progress. In Le Roy's view of history, knowledge was transmitted and lost again; it emerged from the stream of history and disappeared again. The declining phase, however, never ended in total loss and oblivion; some knowledge was always preserved, and in the next phase the successors did not have to start from scratch (Le Roy 1575, 117r-v). For Le Roy, therefore, the expansion of the sciences and arts in the Renaissance was the ascending phase of a new cycle, surpassing the previous ones in the quality and extent of knowledge. The whole development resembled a spiral trajectory (Claussen 2018, 91–92; Werle 2007, 422–424).

All of these models, *prisca sapientia*, the cyclical reappearance of the truth and the spiral advancement, can also be found in interpretations of the history of natural philosophy and mixed sciences. Like in other chapters of this book, I want to substantiate this claim by outlining the basic features of the historiography of astronomy in the Renaissance and early modern periods.

Renaissance astronomers agreed with theologians, philosophers, and physicians that at the beginning of human history, before the Flood, Adam's descendants possessed extensive astronomical knowledge that was lost (Špelda 2013; Steele 2021). Unlike the philosophers, the astronomers did not believe that it was possible to reconstruct the knowledge of the biblical patriarchs. References to ancient antediluvian astronomy, which appeared regularly in the prefaces to astronomical writings, were intended to enhance the dignity and respectability of astronomy because its ancient origins made it part of the long history of Christianity. Astronomy's noble origins with the biblical patriarchs gave it religious and moral legitimacy since traditionally astronomy was condemned as a manifestation of vicious *curiositas*.[5] At the same time, antediluvian astronomy was intended to serve as an orienting ideal that would frame a reform of astronomy. At least rhetorically, the transformation of astronomy could present itself as a return from a corrupted state to the restoration of a more perfect state.

In chapter two, we have seen that Renaissance astronomers were convinced that the astronomy of their time contained many errors, inaccuracies, and unacceptable mathematical constructions. Astronomers such as Regiomontanus and Tycho believed that they were at the beginning of a new phase of flowering in which their discipline would be restored to its former dignity.[6] The ongoing reform of astronomy was mostly understood by astronomers as part of a larger process of alternating phases of *declinationes* and *instaurationes*. The idea of such cycles can be found, for example, in the preface to Johannes Stadius's *Tabulae Bergenses*

[4]Rossi 1970, 67, 76; Mazauric 2007, 189–190; cf. also Delvaille 1910, 142; Bury 1920, 44–43; Baron 1959; Schlobach 1980, 31, 147–148, 202–204; Rapp 1992, 177–179, Rouvillois 2010, 114–115.

[5]Augustine 1981, 178; cf. Blumenberg 1961.

[6]Regiomontanus 1834; Brahe 1913–1929, vol. 5, 9; cf. Swerdlow 1990; Byrne 2006.

(1560). According to Stadius, the history of astronomy had cyclical character: a combination of *translatio studii* and *instaurationes*. Astronomy moved from one part of the world to another, always starting out difficult and slow, but then blossoming again, reaching a peak, and finally moving elsewhere—on a long journey from Assyria to Germany.[7]

The Protestant astronomers perceived cyclical changes in the history of astronomy in accordance with Melanchthon's providential interpretation of history.[8] Under his influence, the Protestant astronomers were convinced that Divine Providence oversaw the history of astronomy and determined its course according to its hidden intentions (Reinhold 1551, fol. 3r-v). According to German astronomers such as Rheticus and Erasmus Reinhold, during the period of decline, astronomy, along with the mathematical disciplines, never completely disappeared because Providence always preserved the mathematical disciplines. The persistence and renewal of astronomy throughout history were therefore not due to humans, but to God, whose caring Providence did not allow the total loss of such a noble and useful discipline.[9] Thanks to Providence, astronomy did not fall into a state of total absence of knowledge at the time of its decline. There was always something left for other astronomers who could resume the interrupted development at the beginning of the next flourishing period (Reinhold 1551, fol. β2v).

Protestant astronomers derived the importance of Providence from the conviction that the history of astronomy was part of eschatology and the history of the world. Astronomy was related to the beginning of the world through the idea of antediluvian astronomy. Similarly, astronomy was linked to the end of the world by the millenarist belief in the flowering of knowledge at the end of history. Some Renaissance astronomers even interpreted the contemporary blossoming of astronomy as the last great cultural upsurge before the end of the world.[10] Astronomy, then, was not an autonomous human endeavour, but a small part of God's overall plan for the whole creation.

In the mid-seventeenth century, the interpretation of the history of astronomy changed. Several surveys of the history of astronomy and mathematical disciplines were published, compiled by Gassendi, Andreas Cellarius, Gerhard Vossius, Cassini, and Claude François Milliet Dechalles, among others.[11] Huygens formulated interesting notes on the development of astronomy in his *Cosmotheoros* (Špelda 2018).

The authors of the histories of mathematical disciplines considered antediluvian astronomy of biblical patriarchs a part of biblical, i. e. sacred, history which was

[7]Stadius 1560, 4, 13, 25; cf. Jardine 1984, 265.

[8]Melanchthon and Peucer 1844, 726–727; Klempt 1960, 27–29; Bottin et al. 1993, 52–58; Kusukawa 1995, 122–144.

[9]Rheticus, letter to Hartmann, 1542, in Rheticus 1968, vol. 3, 46; see also Reinhold 1834, 820.

[10]Rheticus 1542, fol. A4v; Brahe 1913–1929, vol. 2, 312–315; cf. Håkanson 2007.

[11]Gassendi 1658, vol. 5, 363–386; Cellarius 1661, 81–125; Vossius 1650, 127–222; Dechalles 1690; Cassini 1730.

outside the scope of interest of secular historiography. The true history of astronomy began only with the first extant observations appearing in Greek sources.[12] The authors did not believe in any form of highly developed ancient astronomy recalling repeatedly Seneca's words about difficult beginnings. Dechales, for example, wrote simply that the beginnings of astronomy in Greece were crude (*rudia fuerunt haec initia*).[13] In its early days, astronomy was far from perfect, full of errors, and reliant on inaccurate observations.

When historians of mathematical disciplines recounted the subsequent history of astronomy from the Greeks on, they mostly followed the idea of spiral development. Thus, the development of astronomy was repeatedly interrupted by periods of decline, during which astronomy passed between cultures according to the *translatio studii* model.[14] Despite the periods of deterioration, astronomy always maintained continuity. Uninterrupted development no longer resulted from the care of Providence but from human activity, namely observations. In each ascending stage, astronomers did not have to start from scratch but utilized at least some observations from the previous cycle to overcome cultural entropy. Due to the collecting of observations, astronomy gradually improved throughout history. This spiral development was understood by historians of the mathematical sciences as the realm of human diligence, not as part of a divine plan or the result of blind metaphysical necessity governing cycles of flourishing and decay. Waves of barbarism were also seen as the accidental result of human folly. If external disturbing influences had not entered the realm of science, astronomy itself would probably have improved steadily throughout history (Špelda 2015).

By the end of the seventeenth century, the view that the most recent period of spiral history was different from the previous ones had gained ground. Modern astronomy not only knew more than previous periods: it knew more and faster. Indeed, the recent increase in knowledge was far greater than in any previous period. In his survey of the history of the mathematical disciplines, Dechales insisted that 'the mathematical disciplines have grown enormously during the last century and have advanced more than in the last thousand years' (1690, 3b). At the end of his overview of the history of astronomy, Cassini also stated that 'innumerable discoveries had been made in our century, putting astronomy in much better shape than it had been before it began to be taught in Europe' (Cassini 1730, 30).

The speed of the growth of knowledge separated the last epoch from the previous ones and made it something special. Astronomers and historians seemed to believe that by accelerating the accumulation of knowledge, astronomy could escape the periodic appearance of the dark ages. In this way, the progress of mathematical sciences would become irreversible. The possibility of avoiding the next phase of barbarism was further increased by the fact that the modern age had more means of

---

[12] Gassendi 1658, vol. 5, 371; Cellarius 1661, 90b-91a.

[13] Dechalles 1690, 76b; similarly Vossius 1650, 143; Gassendi 1658, vol. 5, 374.

[14] Gassendi 1658, vol. 1, 381–382; Cellarius 1661, 111a-120b; Dechalles 1690, 82b, 90b, 95a.

preserving knowledge. Authors of histories of mathematical disciplines have particularly emphasized book printing, encyclopaedias, and learned societies.

Thus, at the end of the seventeenth century, the history of astronomy was seen as a gradual growth, although interrupted by periodic disasters. Cassini was sure of the direction of progress: 'Astronomy & Geography were gradually perfected.'[15] The reference point for the progress of the mathematical sciences was no longer the past because modern superiority to classical antiquity was obvious to historians of mathematical disciplines. They looked towards the future, to further perfection. The idea of perfection, however, was only vague and indicated neither the goal of progress nor its course. The concept of further perfection consisted mainly of the simple idea of accretion, which was randomly exposed to external catastrophic influences.

The vague idea of perfectibility implied no necessity for progress. Historians of mathematical disciplines represented the history of the sciences mostly as a mere succession of authors and their books. The history of the mathematical sciences, therefore, sometimes resembled more of a list of names and books that were listed in chronological order. Historians no longer worked with the idea of Providence overseeing the history of astronomy and mathematics, but lacked the element that would give meaning, structure, direction, and necessity to the history of knowledge.

Thus belief in reaching perfection did not function as a coherent concept of progress that would consistently link past, present, and future research into a meaningful whole. Fontenelle was able to provide such a concept.

6.2 History of the Human Spirit

In the period from 1683 to 1688 Fontenelle was experiencing his peak and most philosophically creative period (Niderst 1972, 124–126). During this time, Fontenelle wrote several philosophical texts in which he formulated the foundations of his theory of progress.

As early as 1683, Fontenelle published the fictional *Nouveaux Dialogues des Morts* (1683), which hinted at some of his later ideas in an ironic and dialogical form. In an unpublished fragment, *Sur l'histoire*, Fontelle formulated the basic tenets of his philosophy of history. He later reworked the text into a short treatise, *De l'origine des Fables*, which was published in 1724. In 1686, Fontenelle wrote *Histoire des Oracles*, in which he ironically attacked the superstitious belief in oracles. The same year saw the publication of the first edition of *Entretiens sur la Pluralité des Mondes*, followed by dozens of other editions in subsequent decades. In 1688, Fontenelle published another short text, *Digression sur les Anciens et des Modernes*, by means of which he articulated his standpoint in the quarrel of the Ancients and the Moderns.

[15] Cassini 1730, 19; cf. *Mémoires de l'Académie royale des sciences* 1729–1734, vol. 7, 233, 463.

With his philosophical texts, Fontenelle entered the fervent debate on the nature of history and development of human thought. In his recent article, Rob Iliffe noted that early modern British natural philosophers recognized the synchronic existence of several simultaneous temporal frameworks such as biblical history, salvation history, the relationship between the Ancients and the Moderns, techno-scientific progress, belief in the gradual decay of Earth, and the cyclical nature of time (Iliffe 2000). Continental philosophers recognized similar frameworks. The plurality of these frameworks followed from the debates about history that was conducted from different perspectives in the second half of the seventeenth century.

An important theme of discussions and polemics was the binding nature of Christian eschatology as the defining model of history. According to Arno Seifert, the end of the seventeenth century saw the neutralization (*Entkräftung*) of eschatology, and philosophers and historians came up with different interpretations of the beginning and course of human history. These interpretations were based on philosophical reflections on human nature, but also on revived ancient models of human history, such as catastrophic theories, cyclical theories, and geoethnography (Seifert 1986). Another major theme in the dispute was the chronology of biblical history. Astronomers, philosophers, philologists, and natural historians began to comment on biblical chronology from their disciplinary perspectives.[16] The interpretation of history was also influenced by hypothetical conceptions of the natural state formulated by legal theorists and social philosophers (Zedelmaier 2003).

Early modern intellectuals further discussed the direction and course of the history of knowledge: whether it occurred in cycles, or whether there was a decline or, conversely, improvement and progress.[17] The quarrel of the Ancients and the Moderns, which forced proponents of both sides to conceptualize their relationship to the ancient heritage accurately, was important for the historical self-assessment of modern culture, including sciences.[18] The unified vision of sacred history was eroded by reports of New World natives whose cultures came to be seen as preserved remnants of the European past. The 'savages' of America became the starting point for further speculation about the origins of the human race (Perinetti 2006, 1122).

Fontenelle's philosophical importance lay in the fact that he synthetized partial temporal frameworks into the theory of the progress of the human spirit. Although the theory was in many respects receptive and derivative, adopting ideas and views formulated by other authors, Fontenelle was able to bring various views into a coherent whole presented in a concise and attractive literary form. When he became secretary of the Académie in 1697, he began to systematically apply the metaphor of the human spirit and its connotations to the history and advancement of the sciences.

Overall, Fontenelle's conception of the history of human knowledge reflected the radical change in epistemological and anthropological presuppositions that took place in the seventeenth century and that I have partially described in the first two

---

[16] Seguin 2001; Poulouin 1998, 145–186; Rossi 1970.

[17] Schlobach 1980, 194–216; Rouvillois 2010, 110–144.

[18] Duris 2016, 2022; Seguin 2022.

chapters of this book. The Renaissance anthropocentric conception of nature, which was naturally open to the human desire for knowledge, was replaced by a concept of nature that was indifferent to whether humans knew it. For the natural philosophers of the late seventeenth century, the desire for knowledge did not stem from a metaphysical connection between human knowledge and the world that wanted to be known, but from human nature, whose cognitive and psychological features could be grasped and directed to generate scientific progress. For Fontenelle, the source or principle of progress was neither divine nor metaphysical; it originated in the basic features of human nature, such as curiosity.

In his interpretation of the Enlightenment philosophies of history, Dario Perinetti distinguished two main approaches. According to the first approach, the interpretation of history derived from abstract metaphysics of human nature independent of history. The history of humankind could be then characterized as 'the progressive unfolding of innate and uniform natural faculties'. The understanding of history was thus based on knowledge of human nature. According to Perinetti, Jean-Jacques Rousseau, Gotthold Ephraim Lessing, and Immanuel Kant were examples of this line of thought. According to the second standpoint, history was constitutive of the metaphysics of human nature. Human nature was transformed throughout history 'and cannot be understood independently of the exertion of human faculties in history'. This view was typical of Giambattista Vico and Adam Ferguson (Perinetti 2006, 1122).

Fontenelle was a representative of the first attitude. In *De l'origine des Fables*, Fontenelle wrote that he could deduce all history from human nature without speaking of a single specific event (Fontenelle 1989a, 177). In his philosophy of history, he indeed did not speak of historical events but concentrated on the aspects that he believed could be deduced from abstract human nature including epistemic habits.

Similarly to other representatives of this form of the philosophy of history, Fontenelle started by depicting the beginning of human history but he did not use the term state of nature. Unlike other historians and philosophers, he was not interested in the moral, social, or economic aspects of the primordial state. Fontenelle concentrated on the epistemology of the early human race. He sought to reconstruct how early humans thought about the world, reached their conclusions, and communicated their knowledge (Seguin 2017a, 350). He understood his account to apply to all human societies of all places and times. The thinking of early humans, he argued, was universal precisely because it was anthropologically conditioned. All humans had the same nature in relation to knowledge, and therefore all early human societies thought about the world in the same way.

According to Fontenelle, at the beginning of history humans were not noble sages, but 'poor savages' who lived 'in perfect ignorance' and comprehended natural processes as miracles. Their relationship to the world was not based on reason and experience, but on imagination and cognitive passions such as admiration, wonder, and curiosity. As in the *Entretiens*, Fontenelle's short philosophical works did not evaluate wonder as a cognitive passion that brought humans to philosophy, but as a manifestation of the ignorance and backwardness of humans who interpreted the world superstitiously.[19] As Seguin noted, Fontenelle naturalized wonder by placing it in the history of the human spirit as a natural phenomenon of human thought (Seguin 2017b, 280). Wonder was simply a developmental phase corresponding to the weak ability of early humans to rationally conceptualize nature. According to Fontenelle, early humans began to formulate their miraculous explanations of the natural world in the form of fables and myths.

Traditionally, fables and myths were considered evidence from the beginning of human history. Early modern scholars had a long debate about how to interpret ancient fables, which was thoroughly presented by Claudine Poulouin. Many of the participants in this debate believed that there was meaningful content in the fables. Some thought that the fables contained ancient sacred truths originating from God. Others assumed that fables were allegories containing details about the origins of the human race, or information about ancient rulers hidden behind euhemerist narratives. Some early modern scholars understood fables as testimonies of how idolatry gave rise to delusions.[20] All of these interpretations presupposed the ability to understand myths as allegories or cyphers, and therefore required interpreters to be highly erudite in the mythology of classical antiquity (Mullet 2011, 134).

Fontenelle argued that the absence of experience and rational methods necessarily led the first humans to tell stories about natural events, mixing facts with products of their imagination. Gradually the stories turned into a complex system of fables and myths. From an epistemological point of view, fables had a single origin. They were all the product of 'universal mental anthropomorphism' (Hourcade 2003, 196) or a 'universal primitive mentality' (Marchal 1997, 151), because in fables the explanatory principle was mostly anthropomorphic gods and anthropocentric nature (Fontenelle 1989a, 189–190; 1989b, 169–172). For that reason, Fontenelle considered classical mythological erudition with all its interpretive techniques to be a waste of time. There were no hidden truths in fables. Fontenelle stripped myths of their grandeur because he presented them as natural though naive manifestations of human thought at the beginning of history. Poulouin summarized Fontenelle's position using a variation of the reversed criterion of age: mythology was the childhood of reason, not old wisdom (Poulouin 1998, 358).

Fontenelle, however, neither scoffed at the fables, nor considered them an instrument of political manipulation as libertines did. Fontenelle pointed out that although fables seemed like 'chimaeras' and 'absurdities' to the modern reader, their authors and their original audiences found them convincing. This relativization of correctness or truth was a typical feature of Fontenelle's progressivist philosophy of history: the difference between truth and superstition became a difference in the developmental stage of the spirit (Rioux-Beaulne 2012, 350). As Mazauric pointed out, this developmental dynamic separated Fontenelle's concept of fables from the

[19] Fontenelle 1989b, 169–171, cf. Fontenelle 1989a, 187–193, see also Fontenelle 1989–2001, vol. 2, 403–405; cf. Poulouin 1998, 351–358.

[20] See broad overview given by Poulouin 1998, 215–370; see also Bietenholz 1994, 220–284; for the British context see Levitin 2015, 215–229.

libertines of his time. In Fontenelle's view, fables represented the natural way in which early human societies conceptualized the world (Mazauric 2008–2009).

Fontenelle believed in the epistemological and psychological uniformity of early humans in relation to explaining the natural world. He assumed that, given the same social circumstances, humans would think about nature in the same way. As a result, the fables of different societies were similar to each other. Fontenelle remarked succinctly that 'the same ignorance produces almost the same ideas.' Hence, too, there was a great deal of conformity between the fables of the Greeks and those of the 'Americans'. In contrast to the humanist tradition, Fontenelle argued that the early Greeks, for all their spirit, did not think better than the 'Americans'. If the 'Americans' had had enough free time (*loisir*) they would even have thought as rationally as the Greeks (Fontenelle 1989b, 174–175). In *Entretiens*, Fontenelle painted a similar picture of the indigenous peoples of pre-Columbian America first meeting the Europeans who arrived in huge ships armed with firearms. The Indians were terrified because they were at an early stage of epistemological development that had already been surpassed by the Europeans (Fontenelle 1989–2001, vol. 2, 53).

The universality of human nature and the diverse speed of development meant the synchronous existence of societies in different developmental stages. For Fontenelle and his readers, the measure and paradigm of human development was European culture, the history of which served to conceptualize the history of other cultures and people. Although this concept was more typical of later philosophies of history, the roots of the 'simultaneous presence of the non-contemporary' (*Gleichzeitigkeit des Ungleichzeitigen*), as Koselleck and others expressed this trope, can be found as early as in Fontenelle (Koselleck 1975, 391; Nolte 2002).

Since fables were a manifestation of human nature, Fontenelle did not believe in the theory of priestly fraud. To him, fables were a natural mechanism by which humans satisfied their curiosity and desire to understand the world. It was only later, during the history of civilization, that fables were used politically, and above all with the help of another typical feature of human nature: respect for authority.[21] Priests were able to understand the nature of the human mind, and as a result took advantage of fables to deceive and control other people, as Fontenelle detailed in *Histoire des Oracles* (Fontenelle 1989–2001, vol. 2, 53).

Fables became the basis of the first philosophical systems developed by the Greeks. These fictional and speculative systems, according to Fontenelle, gradually lost their persuasiveness over the course of history: 'ignorance has gradually diminished.' Humans saw fewer miracles, created fewer false systems, and wrote less fictional history (Fontenelle 1989a, 201; 1989b, 175). At the end of the *Digression,* Fontenelle optimistically declared that reason was improving and humanity was getting rid of prejudices (Fontenelle 1991, 430).

What puzzled and disturbed Fontenelle was the 'miraculous slowness' of the disappearance of ignorance (Fontenelle 1989b, 175). 'Once our minds are possessed

[21] Rioux-Beaulne 2015, 107–108; cf. Mullet 2011, 136–137; Seguin 2017a.

with an error, it is long before it can be eradicated, if ever.'[22] That was why philosophy had been crude and imperfect for centuries. Fontenelle was convinced that he found the causes explaining the slow rate of decline of ignorance in human nature. The first cause was the interconnectedness of individual ideas ('follies'), which supported each other and formed structured wholes. Fontenelle used the example of anthropomorphic Greek mythology, in which one characteristic of the gods enforced another. For example, the ability to love led the Greek gods to the procreation of mortal children, jealousy, and a preference for one's own offspring. The second reason for the persistence of error was simply 'respect for antiquity'. In conclusion, Fontenelle summed up his distinction in one succinct sentence: The first cause 'extends one folly indefinitely, the second preserves it forever'.[23] The first cause led to the multiplication of errors; the second prevented humans from extricating themselves from errors.

Fontenelle's reflections on human thought at the beginning of history were part of a project to outline 'the history of the human spirit' (Fontenelle 1989b, 182–183). The concept of the human spirit signified the unified experience and ideas of all humans—the totality of human thought as developed in the history. At the same time, human spirit signified the order of human thought, which was universal in Fontenelle's view. Although the human spirit was another metaphor for the unification of experience, it also constituted an essential element of Fontenelle's conception of scientific progress because it made it possible to coherently link epistemology to history, in contrast to earlier metaphors that offered rather vague analogies.

For early modern philosophers, the human spirit was an abstract transhistorical entity that was unchanging (Dagen 1977, 19). For Fontenelle, the human spirit was dynamically changing in a long and complicated development. Therefore, his epistemology was historical. The nature of human thought, the human spirit, should be revealed not by examining the origin of the ideas in the mind, but by studying the historical forms of human intellectual productivity. This is why some modern interpreters consider Fontenelle to have participated in a specifically French tradition of historical epistemology.[24]

According to Fontenelle, there was no point in trying to describe and understand human thought by means of abstract metaphysics and epistemology. In 1702, Fontenelle wrote to Leibniz that 'the nature of the human spirit was incomprehensible to the human spirit.'[25] In the same letter, Fontenelle explained that instead of developing abstract epistemology, humans could empirically investigate intelligible historical products of the spirit and find its scope and its properties.[26] By means of

[22] Fontenelle 1989–2001, vol. 2, 247; English translation in Fontenelle 1753, 95.

[23] Fontenelle 1989a, 195–196; cf. Niderst 1972, 76.

[24] Seguin 2017a, 358; Seguin 2020; Pépin 2012.

[25] Fontenelle, letter to Leibniz, 18 November 1702, Fontenelle 1989–2001, vol. 3, 400; cf. Rioux-Beaulne 2012, 359.

[26] Fontenelle, letter to Leibniz, 18 November 1702, Fontenelle 1989–2001, vol. 3, 401; cf. Fontenelle 1989a, 179–180; cf. Fontenelle 1989–2001, vol. 7, 371–372.

historical research, scholars should grasp the laws regulating the production of ideas and its developmental dynamics. The goal of the history of the human spirit was then to understand human thought by means of its history and to explain the reasons for the current state of knowledge. History was not meant to provide lists of discoveries, but to understand the ways in which the human spirit arrived at them. Fontenellian historical epistemology, then, sought both to uncover the cause of error and to explain the laws of progress of human thought.

In his texts, Fontenelle also described the history of theatre and poetry, which he set in the framework of the progress of the human spirit (Fontenelle 1989–2001, vol. 3, 29–31; vol. 2, 2, 383–385). After becoming secretary of the Académie, he included the research of academicians in this broader epistemological programme. From 1697 onwards, for Fontenelle sciences became the dominant source of evidence for the belief in the progress of the human spirit. As Guilhem Armand (2008–2009, 335) observed, in Fontenelle's view the sciences ensured general progress of human thought. If humans did not have the sciences, each generation would start from scratch, perpetually repeating the same mistakes—like birds caught in the same nets for generations, as Fontenelle wrote in *Nouveaux dialogues des morts* (Fontenelle 1989–2001, vol. 1, 84).

Fontenelle deliberately placed his descriptions of investigations, experiments, and observations made by academicians in a historical context. He often referred to other *HARS* or *MARS* volumes in which earlier similar investigations or theories had appeared, or he placed a given epistemic procedure within a longer historical line of inquiry, e. g. Cassini's attempts to measure the longitude of Marseilles.[27]

The academics themselves wrote descriptive, sober reports of their investigations, usually without any attention to the historical context of their experimental work. Fontenelle, however, purposefully embedded some of these investigations in his narrative of the emancipation of the human spirit from authority and illusions. In the *HARS* volume for the academic year 1702, the chemist Wilhelm Homberg published *Essays de chimie*, where he explained the nature of modern chemistry and its methods but did not discuss its history (*Mémoires de mathématique et de physique* 1702–1742, vol. 4, 33–52). In his section of the same volume, Fontenelle placed Homberg's chemistry within his history of the human spirit, emphasizing the gradualness of its development, its obscure origins and emancipation from false ideas (Fontenelle 1702–1742, vol. 4, 45). Similarly, Varignon published a mathematical text on spirals that focused exclusively on the mechanical and geometrical side of the problem (*Mémoires de mathématique et de physique* 1702–1742, vol. 6, 69–131). Although Varignon made occasional references to ancient mathematicians, he did not otherwise comment on the history of science or the history of the problem. In his *histoire* for the same year, Fontenelle reported on Varignon's research, which for him became primarily an opportunity to integrate ancient and modern geometry

[27] Fontenelle and Godin 1733, vol. 2, 163; cf. Cassini's description in *Mémoires de l'Académie royale des sciences* 1729–1734, vol. 10, 56–62.

in the grand narrative of the progress of the human spirit (Fontenelle 1702–1742, vol. 6, 47–53).

In 1705 La Hire presented his mathematical analyses of the so-called magic squares (*Mémoires de mathématique et de physique* 1702–1742, vol. 7, 127–171, 364–382). Fontenelle again interpreted La Hire's mathematical research as a part of the development of human thought. He attributed the original belief in the magical properties of mathematical constructions to the childlike developmental stage of the human spirit (Fontenelle 1702–1742, vol. 7, 70–81). At other times, Fontenelle was concerned only with sketching the recent history of the problem. In 1707 Guisnée explained the mathematical and mechanical aspects of projectile shooting (*Mémoires de mathématique et de physique* 1702–1742, vol. 9, 140–147). In his part of the same volume, Fontenelle presented the recent history of theories describing projectile motion (Fontenelle 1702–1742, vol. 9, 121–123).

Fontenelle presented each event of scientific research as a unique episode in the development of the human spirit. His texts in the *HARS* volumes, as Seguin observed, thus provided 'a panorama of the advances of the human spirit.'[28] Fontenelle paid special attention to explaining the superiority of modern sciences over ancient ones. More precisely, he sought to define their relationship through the categories of progress and the human spirit. The Ancients represented an important and irreplaceable phase of development, which the Moderns naturally surpassed with their better methods and instruments (Fontenelle 1702–1742, vol. 3, 87, 244; vol. 8, 48, 64). In this historical narrative, scientific research appeared as a process successively pursuing a single goal: the improvement of human knowledge, which included both the accumulation of correct knowledge and emancipation from infantile fantasies.

Fontenelle contrasted the history of the human spirit with political history. At the beginning of *Préface sur l'utilité des mathématiques et de la physique*, Fontenelle formulated a depiction of 'perpetual revolutions in human affairs, of the rise and fall of Empires' (Fontenelle 1994, 45), which is according to Poulouin reminiscent of Jacques-Bénigne Bossuet's *avant-propos* to his philosophy of history. While Bossuet highlighted the role of Providence in 'the spectacle of perpetual revolutions';[29] in Fontenelle's universe, history was merely an unfortunate display of human passions and follies that did not lead to the realization of any meaningful goal. The history of the human spirit, by contrast, had meaning, inner order, and purpose. Fontenelle argued that in this history, scholars could notice a gradual increase in comprehension of the order of nature, including its Creator (Fontenelle 1994, 45–47). Hence the history of the human spirit represented a much nobler subject than the ordinary history of human deeds. 'One of the most pleasing Histories, and, without doubt, most Philosophical too, is that of the Progress of Humane Minds.'[30]

---

[28] Seguin 2016b, 324; cf. Seguin 2012; Rouvillois 2010, 147; Schlobach 2001, 1002; Marsak 1959, 3, 49.

[29] Bossuet 1836, 321–322; cf. Poulouin 2006, 75.

[30] Fontenelle 1702–1742, vol. 6, 172; English translation Fontenelle 1727, 103.

Political and social history was a showcase of folly and futility, while the history of the spirit represented an intelligible process that had the value of allowing humans to understand the patterns of human thought and orient future epistemological efforts accordingly. Fontenelle, therefore, did not want to write a triumphalist picture of the progress of the sciences. In his view, the human spirit had a strong and natural propensity for error that needed to be properly studied and grasped.

That is why history should study all human errors and to critically follow the spirit 'in all the stages by which it has reached the last excesses of absurdity' (Fontenelle 1989b, 182). In the *éloge* of Leibniz, Fontenelle noted that the history of human thought was always instructive because it 'shows the greatest geniuses that there were others like them and that they too were mistaken' (Fontenelle 1989–2001, vol. 6, 390). The negative exemplarity of predecessors, as Rioux-Beaulne remarked, allowed contemporary scholars to understand the general principles behind the fallibility of the human spirit (Rioux-Beaulne 2012, 361). Knowledge of these principles was meant to help natural philosophers avoid the mistakes that their predecessors had made in the past.[31]

Fontenelle's outline of the progress of the human spirit thus always embraced two sides. The first was positive and celebratory. Its content was a historical depiction of the cumulative gathering of knowledge and inductive discovering laws (Fontenelle 1989–2001, vol. 7, 337–343). The other side was negative because it narrated the story of progress from the opposite perspective, as the diminishing of ignorance. The dynamic factor of negative progress was emancipation from prejudice and dogma. The combination of the positive and negative sides of progress produced the adventurous story of the emancipation of the human spirit, which became the dominant interpretive framework for the history of human thought during the Enlightenment (Dagen 1977, 59–66; Israel 2004).

6.3 Periodization of History

Fontenelle emulated the older tradition of metaphorically linking history with the ages of human life. Both, the development of universal man and the collective human spirit repeated the cognitive development of the individual. Phylogeny repeated ontogeny. The advancement of human knowledge proceeded in the same stages as the coming of age of a single human from naive infancy to rational maturity. Fontenelle mostly worked with these two extreme positions of development: early childishness and late maturity. He did not utilize the more detailed association of stages of human life (*pueritia* or *adolescentia*).

The developmental division of the progress of the human spirit into the stages of infancy and maturity was not convincing in itself. Fontenelle therefore combined

[31] Fontenelle 1989a, 201; 1989b, 179; 1989–2001, vol. 7, 359.

epistemological development with an elementary chronology that made it possible to locate the emancipatory and cumulative aspect of the development of the spirit in history. Such an epistemologically based periodization of history provided a comparison of the epistemological level of different epochs that could serve to evaluate their level of progress. Finally, an important part of this periodization was the self-definition of the modern age. In Fontenelle's narrative, the modern period and its knowledge should appear as the logical and necessary culmination of previous developments. In this way, Fontenelle wanted to refute the objection of the defenders of the Ancients that the achievements of the modern sciences were merely a contingent phenomenon of history that would disappear again after a while.

Fontenelle never formulated his periodization of the history of the spirit as clearly as his philosophy of history, but we can reconstruct it retrospectively from references that appear sporadically in his philosophical texts or the *HARS* volumes.[32]

I agree with Mazauric that Fontenelle was working with a chronological scheme that had five phases (Mazauric 2007, 39). Before the beginning of history, there was an ahistorical period derived from the metaphysics of human nature, which applied universally to all humans. The second phase was the classical ancient world. In Fontenelle's view, the Greeks and Romans had achieved an imperfect knowledge of nature based on limited experience. They experimented little and were subject to prejudice and dogma.

After the demise of classical antiquity came a period of decline. The traces of ancient knowledge were almost lost and 'the long Darkness of the barbarous Ages' began. In the Middle Ages, natural philosophy was 'sterile' and made no advancement for several centuries (Fontenelle 1989–2001, vol. 7, 337). The fourth phase was the Renaissance, bringing a great revival of the sciences. However, the development of the spirit progressed very slowly in this period because natural philosophers and natural historians depended too much on the authority of ancient texts instead of investigating nature itself using empirical methods.

According to Fontenelle, the modern period had started about a hundred years prior. 'Philosophy shook off the yoke of authority' and began to follow the guidance of reason (Fontenelle 1702–1742, vol. 1, 17). Philosophers broke away from the past, shed their reverence for ancient models, and began, as it were, anew. The sciences were reborn. The progress of the sciences, 'which had almost completely stopped, began again with renewed vigour' (Fontenelle 1989–2001, vol. 7, 337). After the new start, the sciences had to struggle with prejudice, authority, ignorance, and disinterest, and yet they achieved outstanding successes in a short time, according to Fontenelle (1989–2001, vol. 7, 339).

The rapid progress of the new epoch was mainly due to a new way of thinking (Fontenelle 1702–1742, vol. 2, 72). Natural philosophy began to investigate nature with a new method that acted as an accelerator of discoveries. Fontenelle likened the method to a 'fountainhead'. To emphasize its importance, Fontenelle seemed to deliberately exaggerate: Applying the method 'one has nothing more to do, but here

[32] Fontenelle 1702–1742, vol. 2, 71; 1989–2001, vol. 7, 357–362 and vol. 3, 239; 1994, 48.

let one self be carried down the gentle Stream of Consequences.'[33] Being simpler, the new methods were better applicable than the Greek ones. As a result, in the empirical sciences, the amount of experiential knowledge expanded considerably. In the mathematical sciences, too, progress consisted in knowing better methods. Modern geometers knew many more truths than the Ancients, not because they had a better spirit but because of their excellent methods (Fontenelle 1702–1742, vol. 6, 53; vol. 3, 87). Progress of the sciences also accelerated as *savants* began to share their knowledge and the first learned societies were formed (Fontenelle 1989–2001, vol. 7, 338–341).

Between the first four periods and the fifth, there was a historical gap represented by Descartes and his philosophy. Fontenelle was not a Cartesian in ontology and refused to subscribe to the strict mechanisms or to make Descartes the new authority (Fontenelle 1991, 430). He saw in Descartes above all the advent of a new way of thinking that was the driving force behind the achievements of modern sciences. While Descartes's philosophy was in many ways mistaken and quite uncertain, his new method was much better (Fontenelle 1991, 420). Embracing *l'esprit géométrique*, the Cartesian method brought order, clarity, precision, and exactitude into science. Fontenelle highlighted that the geometrical spirit spread from geometry to other disciplines, and thus contributed greatly to accelerating their progress (Fontenelle 1994, 44). In physics, Descartes's thought provided an 'infinity of views', some of which proved to be correct. Others, despite being wrong, gave rise to other theories (Fontenelle 1989–2001, vol. 7, 339). For instance, Huygens, according to Fontenelle, was able to develop his theory of light 'with the help of Descartes's errors' (Fontenelle and Godin 1733, vol. 1, 283).

Descartes's method represented a moment of epistemological epiphany and intellectual maturity in the history of the spirit: 'one opened the eyes and dared to think' (Fontenelle 1989–2001, vol. 3, 239). In the *éloges* of some academicians, Fontenelle represented the encounter with Descartes's philosophy as a moment of enlightenment or conversion, as Mazauric observed (2003, 89–90). Descartes's philosophy freed academicians such as Tournefort, François Poupart and Malebranche from the thrall of scholasticism and led them onto the path of reason (Fontenelle 1989–2001, vol. 6, 189, 217, 338–339).

In the period after Descartes, the advancement of sciences accelerated significantly. In *Préface sur l'utilité des mathématiques et de la physique*, Fontenelle wrote enthusiastically of sciences progressing quickly after a long barbarism: 'we should be astonish'd at the Greatness and Swiftness of their Progress (*la rapidité du progrès des sciences*), seeing so many new ones produced, as it were, out of nothing.'[34] In the *éloge* of Pierre Remond de Montmort from 1719, Fontenelle wrote that on the route of the human spirit, the intervals were at first very long, but 'naturally' became shorter.[35] By emphasizing the natural character of acceleration, Fontenelle wanted to

[33] Fontenelle 1989–2001, vol. 6, 103; English translation in Fontenelle 1727, 32.

[34] Fontenelle 1994, 47; English translation in Fontenelle 1727, xvi.

[35] Fontenelle 1989–2001, vol. 6, 475–476; cf. Mazauric 2007, 262.

articulate the necessity of this process. Necessary acceleration elevated the present epoch above all previous ones, minimizing its contingency and preventing its achievements from being lost. When Fontenelle and his colleagues used the hyperbolic assurance that the sciences in their age were making discoveries 'every day', they meant to imply that the progress of the sciences was irreversible. By sustaining and accelerating itself, scientific progress achieved sufficient escape velocity to protect the sciences from oblivion.

The historical scheme of the five periods presented the background to Fontenelle's reflections on the history of the sciences. Fontenelle used it implicitly as a general framework when turning to the history of a particular field or explaining a current scientific problem. In this way, he could introduce partial research results as parts of a long chain of knowledge that had its own history and origin. From the point of view of contemporary readers, this period scheme fitted well into the periodization of the intellectual history with which they were acquainted. The negative view of the Middle Ages and the Renaissance was in line with the belief in the existence of oscillations that caused alternating phases with different levels of scholarship. Fontenelle, however, did not address the question of how historical changes occurred and how one period came into being and another faded away. It seems that behind the changes and transitions from one period to another were *révolutions* (Fontenelle 1702–1742, vol. 1, 118; Fontenelle and Godin 1733, vol. 1, 41), which he seemed to understand in a secular reinterpretation of Bossuet's philosophy of history as large and contingent political changes.

Fontenelle's view of progressing human spirit went beyond the narrow quantitative framework of the descriptive idea of progress based on the simple idea of accretion. Fontenelle created a historical narrative that gave origin and meaning to human history by explaining it as an epistemological project that had negative (emancipatory) and positive (accumulation and induction) sides. The sciences played a crucial role in this process as they were a means of emancipation, understanding, and intellectual maturing.

6.4 The Order of Progress

In Fontenelle's view, the history of human thought did not happen randomly, but lawfully and according to a 'necessary relationship' (Fontenelle 1989b, 184). Unlike irrational political history, the history of human thought was subject to rules that enabled philosophers to reconstruct its past and, to some extent, predict its future. Fontenelle never explained these rules systematically. Sometimes he hinted at them in his texts, and sometimes they were present as tacit assumptions. The list of rules that follows is thus an attempted reconstruction.

The first rule of the progress of the spirit may be defined as directionality towards general knowledge. According to Fontenelle, natural philosophers collected empirical knowledge, found the necessary connections using reason, discovered partial regularities, and finally formulated general principles (Fontenelle and Godin 1733, vol. 1, 20, similarly Fontenelle 1702–1742, vol. 4, 118; vol. 8, 74, 100). In Fontenelle's view each particular piece of knowledge comprised a link in a chain leading to an inductive conclusion. Hence, the history of the spirit was a gradual historical induction. Beginning with disjointed facts and complicated anthropomorphic modes of explanation, human knowledge became increasingly simplified. Fontenelle expected this process to culminate in the formulation of few simple general laws.[36] The gradual inductive derivation of general principles served the human spirit as a means of coping with the infinite variety of nature. Human thought could not grasp the 'infinity of facts' and therefore had to reduce them to 'a small number of general principles which contain, as in abbreviation (*comme en abregé*), particular facts'.[37]

Knowledge of general principles was the main and most important tool of human thought. Fontenelle never hoped that humans would achieve absolute knowledge of the natural world. Humans were not able to grasp the particulars as God does; therefore, they would remain dependent on nominalistically conceived general concepts that served as abbreviating ways of the spirit (*voie abrégée de l'esprit*).[38] But this was precisely what confirmed the autonomy of human knowledge. Deriving empirical laws independently, the human spirit owed its knowledge to no one, and this gave Fontenelle and his contemporaries pride, self-confidence, and belief in the further development of the sciences. An important part of modern European thought was born: Thanks to the autonomous progress of their knowledge, philosophers and scientists could start to consider themselves as makers of the future.

Another rule of progress was the cumulation of knowledge over time. This was, of course, a key feature of the descriptive idea of progress. After comparing the two epochs, it was easily discernible that the later epoch knew more than its predecessors. Fontenelle explained this rule simply: Ptolemy lived later than Hipparchus and therefore had more observations and could correct some of Hipparchus's inaccuracies 'owing to the natural advantage which the later always have in these matters' (Fontenelle and Godin 1733, vol. 1, 41). In modern times, Torricelli knew more than Galileo 'because he came after him' (Fontenelle 1702–1742, vol. 9, 122). Descartes introduced the first theory of the Earth, but after him came Nicolas Steno, Thomas Burnet, John Woodward and Johann J. Scheuchzer, who corrected his ideas and built on each other (Fontenelle 1702–1742, vol. 10, 30).

Fontenelle did not stop at merely noting the cumulation of sciences in his reflections. He went beyond the descriptive idea of progress because he was aware of the relativistic mechanism of the temporal regime of knowledge. The history of sciences did not consist of merely adding new knowledge to old. In fact, new discoveries sparked further unexpected discoveries: 'the mind has already been enlightened by the very discoveries one has before one's eyes.'[39]

---

[36] Beugnot 1989, 493; Seguin 2012, 377; 2023, 193.

[37] Fontenelle and Godin 1733, vol. 1, 162; cf. Niderst 1972, 214–215.

[38] Fontenelle 1989–2001, vol. 7, 477; cf. Mullet 2011, 122–123; Mullet 2012; Niderst 2002.

[39] Fontenelle 1991, 418; English translation in Fontenelle 1960, 361.

The cumulation of knowledge intensified when there was direct continuity between scientists. According to Fontenelle, the English mathematician John Wallis surpassed Bonaventura Cavalieri because he 'came after the Italian' and began where the Italian mathematician left off (Fontenelle 1989–2001, vol. 7, 461–462). Descartes picked up where the Ancients left off, and Leibniz started up where Isaac Barrow finished (Fontenelle 1989–2001, vol. 3, 239, 241). In Fontenelle's view, unlike classical antiquity and the Renaissance, the modern age brought about a seamless continuity between scientists from different countries (Mullet 2011, 186). Hence a temporal and geographical condensation of research occurred that was simultaneously the result of progress achieved and the cause of further acceleration of development.

The cumulative nature of knowledge was also manifested in another way. Fontenelle believed in the natural interdependence of true knowledge: 'all truths are enlightened by each other'.[40] From the interconnectedness of truths followed the fruitfulness of truths because Fontenelle assumed that each piece of knowledge discovered could be a path to the next: 'it is the essence of truth to be fruitful' (Fontenelle 1702–1742, vol. 5, 6; vol. 6, 89). For instance, the tiny parallax of Mars revealed by the astronomical expedition to South America helped astronomers calculate the distance of Mars and the immeasurable size of the Sun (Fontenelle and Godin 1733, vol. 1, 170). Therefore, Fontenelle argued, academicians should pay attention even to seemingly useless observations and experiments. It was always worthwhile to gather empirical knowledge, even if philosophers were not sure of its meaning: 'let what will come of it, the Risque cannot be great.'[41] In Fontenelle's view, progress did not always have to be guided by a sophisticated scientific method. Sometimes it was enough to accumulate a large amount of observations in hope that the sheer volume of data collected could generate further progress.

The third rule of progress was continuous linearity. Fontenelle regarded the progress of the human spirit as lawful because he saw it as an autonomous process that sustained itself. New discoveries and methods were like a 'fountainhead' from which more discoveries flowed. In today's terminology, then, Fontenelle's concept could be described as that of an internalist. At the beginning of the history, humans explanations of nature were necessarily full of errors and infantile prejudices. Then, in the sphere of pure knowledge, constant progress should follow but it was hampered, limited, and overshadowed throughout history by external influences such as war, superstition, ignorance, and the neglect of education. Typically, the barbarism and horrors of war in the Middle Ages 'suffocated the sciences for a long time' (Fontenelle 1989–2001, vol. 3, 29).

The emergence of the 'dark ages', as Rouvillois remarked, was neither natural nor necessary part of the mechanism of history, but an external disturbance of a process that would otherwise have been continuous (Rouvillois 2010, 140). Fontenelle, however, lacked the conceptual means to explain the disruption of the progress of

[40] Fontenelle 1994, 43; English translation in Fontenelle 1727, x.

[41] Fontenelle 1994, 43, English translation in Fontenelle 1727, ix.

the spirit, and therefore resorted to metaphors. In *Digression*, Fontenelle presented the Middle Ages as a time of forgetting, 'an illness, a passing amnesia', after which the human spirit could educate itself again (Schlobach 2001, 1101).

The linear model of the development of knowledge that Fontenelle, and probably his colleagues at the Académie, advocated was not nearly as natural at this time as it might seem to today's reader. In the late 1680s and early 1690s, the first French phase of the *querelle des Anciens et des Modernes* was underway, which revealed, among other things, that contemporaries with literary educations preferred a descending interpretation of the history of knowledge or an oscillation of blossom and decline.

In the four volumes of his *Parallèle des Anciens et des Modernes,* Charles Perrault made a meticulous comparison of the achievements of the Ancients and the Moderns in the sciences and arts. Perrault's view of the intellectual history was guided by the descriptive idea of progress, and he was therefore primarily interested in demonstrating the superiority of the present over classical antiquity. He was so impressed with the advance of knowledge in the recent past that he was almost 'incapable of imagining further progression,' observed Bury (1920, 87). Perrault indeed did not outline any perspective of future progress. He was convinced that knowledge had reached its peak in his century and that only slowing progress and decline would follow (Perrault 1688–1697, vol. 1, 98).

The defenders of the Ancients mostly recognized the superiority of the Moderns in the sciences. But they relativized the exceptionality of the modern age. In their view, the modern scholars were no better than the Greeks, and their age did not represent the peak period of scientific inquiry. Abbé Jean-Baptiste Dubos, for example, admitted that 'the natural sciences were more perfect today', but this did not mean that modern scholars could think better (*mieux raisonner*) (Dubos 1719, vol. 2, 426).

When Fontenelle introduced the idea of linear and cumulative scientific progress open to the future, it was a novelty that was surprising to many contemporaries and contradicted their intuitions and education. The introduction of the prescriptive idea of the progress of the sciences marked as profound a change in thinking about nature as the rejection of geocentrism or the idea that nature had a history.

The quarrel of the Ancients and the Moderns also revealed another consideration that was important for the constitution of the idea of scientific progress: its polychrony. Participants in the quarrel continuously used the division of disciplines into two groups. The first group consisted of disciplines based on reason or another faculty of mind (e. g. imagination or genius). These 'intellectual' disciplines depended solely on the power of mind, genius, or talent. They included poetry, eloquence, literature, drama, painting, sculpture, and other arts. The second group was comprised of disciplines focused on the empirical investigation of nature, i.e. the sciences. Such a division can be found in various participants in the quarrel such as in Hilaire-Bernard de Longepierre, François de Callières, Abbé Dubos, William Wotton, Georg Pasch, and Georg H. Ayrer.[42]

The main difference between the two sorts of disciplines was their different historical development. 'Intellectual' disciplines, depending solely on the power of the mind, reached perfection very soon after their inception. Empirical disciplines based on a gradual accumulation of data and discoveries were considered to run very slowly. Thus, 'intellectual' disciplines could have reached a high level of development very early in history, while empirical sciences needed a long time to accumulate experience.

Participants in the quarrel agreed that the different pace of development was also a source of the historical discrepancy between the two groups of disciplines. 'Intellectual' disciplines had been perfected at the beginning of history. Empirical disciplines matured in the modern period. In *Digression*, Fontenelle summarized the debate by saying that rhetoric and poetry depended on the 'liveliness of the imagination' and required neither method nor long-term accumulation of experience. That was why they also came to a peak relatively early. But empirical sciences, medicine, and mathematics, were composed of numberless ideas and depended upon a precision of thought that improved 'with extreme slowness'.[43] In the *éloge* of Malebranche, Fontenelle, in an impressive abridgement, said that genius made philosophers and poets, and time made scientists (*savants*) (Fontenelle 1989–2001, vol. 6, 339).

Thus, Fontenelle and his contemporaries assumed that there was a difference in the rate of progress between 'intellectual' arts and empirical sciences (including mathematical disciplines). In his capacity as secretary, Fontenelle found that there were also differences between the individual empirical sciences. In *Préface sur l'utilité des mathématiques et de la physique*, Fontenelle claimed, in metaphorical shorthand, that all the sciences were only in their cradles (Fontenelle 1994, 49). He did so, presumably, to temper the exaggerated expectations of utilitarian royal administrators of the Académie. In other texts, however, he highlighted the long history of certain disciplines, such as astronomy and medicine. By contrast, he argued that optics had barely been born and important discoveries were yet to be made, given the 'rapid course of sciences in this age' (Fontenelle 1702–1742, vol. 2, 134). The different sciences were thus at different stages of their development, which, similar to individual biographies, copied the overall development of the spirit. As Pépin remarked, Fontenelle should not be considered a theorist of scientific progress, but a theorist of the progresses of the sciences (*des progrès des sciences*), because each discipline had its own modality of progression (Pépin 2012, 392).

The fourth rule ordering the progress of the spirit was the necessity of its course. Historians considering the general idea of progress, such as Bury, Delvaille, and

[42] Longepierre 1687, 44; Callières 1688, 293, 299; Dubos 1719, vol. 2, 433, 522; Wotton 1694, 19, 78; Pasch 1700, 3–4; Ayrer 1735, 360, 388–389, 393–394.

[43] Fontenelle 1991, 419, English translation in Fontenelle 1960, 362.

especially Rouvillois in his penetrating analysis, have drawn attention to this point.[44] The descriptive idea of progress asserted the superiority of the most recent generation but did not indicate a necessity for further progress. The comparison was always made retrospectively. That is why the natural philosophers and scientists of Fontenelle's time so often cited Seneca's words that knowledge would only continue to expand if the diligence of posterity did not cease. Having no philosophical justification for the continuation of progress, philosophers could only hope for the diligence and responsibility of their successors.

Fontenelle justified the necessity of further advancement of the sciences by demonstrating that this process did not depend on chance, the eschatological plan of Providence, unforeseeable external circumstances, nor on the will of future generations. Progress was manifestation of a purposeful effort springing from human nature unless external influences disturbed its 'natural' course.

In the *Digression*, Fontenelle explained the necessity of progress utilizing his concept of nature and human nature. In Fontenelle's view, nature was neither mysterious nor unpredictable. Nature was stable and invariant therefore the same causes always produced the same effects. Some of Fontenelle's contemporaries, especially the advocates of the Ancients, held a vitalist conception of nature that explained nature as an ageing organism (Duris 2016, 106–145; Duris 2017). According to them, in the times of the Greeks, nature was young and fertile, but in the present, it had lost its productive powers. The poet Longepierre argued that 'nature had exhausted itself' and no longer produced great persons comparable to those of classical antiquity (Longepierre 1687, 203). Similarly, according to Bishop Pierre-Daniel Huet, in classical antiquity, nature was not as depleted as it was in his times, and produced humans who were bigger, stronger, and better, who had more spirit and 'heads better laid out' (Huet 1722, 32).

Right at the beginning of the *Digression*, Fontenelle rejected the thesis that 'nature had exhausted itself' and was no longer producing great geniuses. Nature produced the same trees and human brains over and over again (Fontenelle 1991, 414). The uniformity of nature implied that humans were cognitively and physiologically the same. Neither in history nor geographically did human nature change. Humans still naturally craved knowledge and just as naturally succumbed to error and prejudice: 'centuries do not put any natural differences among men'[45] According to Fontenelle, it was not true that the Ancients had 'much more spirit than us' because they were in relation to the Moderns 'perfectly equal'. In terms of physiological and cognitive dispositions, there was a natural equality (*l'egalité naturelle*) among humans.[46]

---

[44]Rouvillois 2010, 147–166; cf. Bury 1920, 109–110; Delvaille 1910, 215–217, Dagen 1977, 44–46.

[45]Fontenelle 1991, 416; English translation in Fontenelle 1960, 360, cf. Fontenelle 1989–2001, vol. 3, 238.

[46]Fontenelle 1991, 416–417; see comments in Dagen 1977, 37–47,123–127; Marchal 1997, 153–158; Mullet 2011, 131–134.

As a result of the passage of time and cumulative nature of empirical knowledge, an epistemological inequality developed between humans of different centuries relating to the extent of knowledge and its correctness. Throughout history, the extent of the sciences increased and the number of errors and prejudices decreased. Therefore, humans who came later had to know more than those before them. If the postulates of the uniformity of nature and the equality of humans were valid, the epistemological superiority of the Moderns was inevitable. Otherwise, according to Fontenelle, the Moderns would simply not be the same humans as the Ancients. If the Moderns had not expanded their knowledge compared to the Ancients, it would mean that they were cognitively or physiologically inferior: 'if we only equalled them we should be inferior to them by nature.'[47] The progress of the spirit thus 'participates in human nature', as Rouvillois noted in commenting on this passage.[48]

The whole argument was of course based on the validity of the metaphysical postulates of uniformity and equality, but it also relied on other implicit assumptions. The Moderns were to surpass the Ancients only if other rules of progress were kept: cumulation, succession, and continuity in the transmission of knowledge. If later philosophers had to start over from scratch, they would invent the same as the earlier ones.

The fifth rule regulating the progress of the spirit was its lawful course. Progress did not proceed haphazardly; it had an order. According to Fontenelle, individual truths (*verités*) formed a large coherent whole and followed one another in a certain order, which Fontenelle characterized as a 'necessary succession of discoveries' (Fontenelle 1989–2001, vol. 3, 238). In various texts, Fontenelle described this succession as a consecutiveness of stages or steps, each successive one building on the previous one. In relation to the history of geometry, Fontenelle wrote that the ancient geometers were 'afraid to take one extra step' that would lead them to the idea of geometric infinity (Fontenelle 1989–2001, vol. 7, 359). In the modern age, Leibniz and Newton discovered calculus, and thus 'opened the way' for further discoveries. Along this path, the great geometers of the Académie 'walked with the step of giants', so that geometry was 'allmost totally revolutionized' (Fontenelle 1989–2001, vol. 7, 362).

Fontenelle concluded his interpretation of the history of geometry with the oft-quoted summary: 'There is an order that governs our progress. Each piece of knowledge develops only after a certain number of previous ones have developed, and when its time has come to bloom' (Fontenelle 1989–2001, vol. 7, 364). Each area of knowledge had to go through a sequence of steps that led from partial findings and infantile theories to general principles. Rouvillois observed that the process of cognition was given meaning and a stable structure by the introduction of the 'order of progress', which was made up of successive stages that could only be passed through one by one (Rouvillois 2010, 155).

---

[47] Fontenelle 1991, 419; English translation in Fontenelle 1727, 362.

[48] Rouvillois 2010, 148, cf. Schröder 1990, 986–987.

6.5 The Fatality and Relativity of the Progress of the Human Spirit

The 'order of progress' had non-trivial consequences for understanding the history of sciences and humanity because it organized the history of the sciences into an unchanging pattern of inductive developmental logic. Philosophers advanced from the empirical gathering of particular findings to general principles, from infantile superstition to the laws of nature. The necessity of this process resulted in the idea of knowledge distributed throughout history according to its epistemological quality. Singular and isolated knowledge appeared at the beginning, because the human spirit had to wait for general truths and laws. In Fontenelle's view, the most natural ideas were the simplest and most general. They embodied the ultimate form of knowledge. These general ideas could not be discovered at the beginning; humans had to wait for them for a long time: 'It is a Kind of Fatality, that in all Arts the Methods, or the most natural Ideas, are not those that offer themselves most naturally.'[49]

The 'fatality' mentioned by Fontenelle signified that humans in certain historical periods could not extend their knowledge beyond the scope given by the succession of stages. Thus, discoveries could not occur at any time but only when the conditions of their possibility in the temporal sequence of the advancement of the spirit arose. In the history of the sciences, therefore, no one could overtake another. Fontenelle summarized this radical idea in a simple simile. If Archimedes had lived at the beginning of the world, he would have invented at most the plough. 'Archimedes living in another century burns the Roman fleet with mirrors' (Fontenelle 1991, 417). In another text, Fontenelle noted that a strange philosopher from the beginning of human history could well be the Descartes of our time (Fontenelle 1989a, 189).

By the imaginative displacement of the scientist in time, Fontenelle meant to suggest that even a brilliant inventor cannot upset the logic of progress. The genius was trapped in her own time (Rouvillois 2010, 156; Armand 2012). This regularity, according to Fontenelle, applied to fools as well. Everyone had to think the way their century thought. The uneducated person in the learned century (*savant siècle*) could not avoid the true knowledge of his contemporaries. Thus not even stupid persons escaped the logic of progress: they were also imprisoned in the light of contemporary sciences (Fontenelle 1989–2001, vol. 3, 62).

The restraint of knowledge by developmental phases did not concern solely individuals, but also generations and epochs. In his preface to L'Hôpital 's *Analyse des infiniment petits* (1696), Fontenelle wrote clearly: 'It does not seem that the Ancients were able to do more in their time; they did what our minds (*nos bons esprits*) would have done in their place; and if they were in our place, it is to be believed that they would have the same opinions as we have. All this is a consequence of the natural equality of minds and of the necessary succession of discoveries' (Fontenelle 1989–2001, vol. 3, 238). The order of the progress of the human

[49] Fontenelle 1989–2001, vol. 6, 99–100; English translation in Fontenelle 1727, 27.

spirit thus reduced the differences in human thought to a difference in position on the timeline.

Posing the inevitable succession of stages, the order of progress undermined exaggerated confidence in human reason and thus became a natural complement to the epistemic practice of sage pyrrhonism. Although Fontenelle was undoubtedly developing the impulses of contemporary French scepticism, he did not agree with the sceptics who denied the possibility of gaining knowledge. Following Gassendi, Fontenelle separated from the sceptic position with the very idea of scientific progress, which comprised gradual neutralizing of errors and slow working on the discovery of general truths. In the great arc of history, the weakness of reason could be compensated for by long-term collective research and patience.

One of the main enemies of this patience was human self love, which Fontenelle portrayed in two ways. The first was the anthropocentric tendency of humans to pass off their speculations as truth and reality. In the *Entretiens*, for example, Fontenelle referred to the anthropocentric cosmology of the ancient Chaldeans and to the delusions of grandeur typical for the Romans, who presumed that the map of the world was not much bigger than the map of their empire (Fontenelle 1989–2001, vol. 2, 23–24, 54–55).

The second form of self-love was an unjustified belief in human cognitive abilities: 'our Vanity often flatter us with the Hopes of attaining the Knowledge of things that are not made for us.'[50] Above all, Fontenelle distrusted reason left to itself. Fontenelle considered human reason to be a tool that should be used with great care since reason and error were mutually attracted: 'The human mind and the false sympathize extremely' (Fontenelle 1989–2001, vol. 1, 68). In his early texts, Fontenelle considered this attraction to be a universal part of human nature: 'Mankind has, in all Ages, the same Inclinations, over which Reason has not the least Power. So that to the World's End there will be Follies (*sottises*) and the same Follies too, as long as there are Men.'[51]

Fontenelle therefore did not share the belief of metaphysicians in grasping the true fundaments of reality. In his correspondence with Leibniz, Fontenelle politely articulated his reticence about the possibility of explaining the world by means of a rational metaphysical system such as that presented by Leibniz. Rather than comprehending Leibniz's metaphysical physics as an significant extension of knowledge, Fontenelle merely accorded it 'sublime beauty'.[52] Later, in the *éloge* of Leibniz, Fontenelle similarly noted that knowledge such as Leibniz's metaphysics was not meant for humans, 'it is too sublime for us' (Fontenelle 1989–2001, vol. 6, 404).

Another folly of reason consisted of an excessive haste to formulate general conclusions and systems. In this view Fontenelle followed other early modern

[50] Fontenelle 1994, 48; English translation in Fontenelle 1727, xv.

[51] Fontenelle 1989–2001, vol. 1, 84; English translation Fontenelle 1730, 91.

[52] Fontenelle, letter to Leibniz, 24 March 1703, Fontenelle 1989–2001, vol. 3, 403–405; cf. also further letter to Leibniz, 6 July 1703, 410; cf. Niderst 1972, 245–247.

natural philosophers who also feared the premature finalization of knowledge into a system that would sterilize further research. The order of progress, however, allowed Fontenelle to better conceptualize the prohibition against haste in drawing general conclusions.

As ontogeny and phylogeny were consistently intertwined in Fontenelle's theory, the human spirit's urge to rush operated at both the historical and epistemological levels. Already at the beginning of history, the first humans formed hasty philosophical systems (Fontenelle 1989a, 188; 1989b, 170). First systems had to be wrong as we already know from the order of progress. On the epistemological level, the danger of haste threatened especially the first discoverers. Fontenelle required them to become aware of their position in the order of progress. According to this order, all early discoveries and theories were erroneous or restricted by a lack of experience and the absence of correct methods. The first discoverers paid for their initial boldness with the absence of truth.[53] Fontenelle described the constraint of initial theories as almost a fatal necessity. Even Galileo was not spared 'this destiny which is common to all great geniuses'.[54] Occasionally, Fontenelle wrote of the bad luck (*malheur*) of those who worked on something first (Fontenelle and Godin 1733, vol. 1, 213).

In reflecting on imperfect beginnings, Fontenelle became aware of the relativistic mechanism of progress that continuously changed the value of discoveries and difficulty of theories. He observed that with the development of new methods, old truths appeared to be very easy to discover.[55] Modern mathematicians would not highly regard the discovery that made Archimedes cry out 'Eureka'.[56] On the other hand, the modern state of the sciences meant that further discoveries were more difficult to achieve. Fontenelle argued that for the first discoveries, the human spirit did not need to exert great effort. Much more effort was needed to achieve further discoveries 'because the subject is more nearly exhausted and what remains in it to be discovered is less apparent to the naked eye'.[57] Advancement was not simply the addition of new discoveries, because progress itself relativized their importance and made it more difficult to achieve further results.

These examples should reveal that by introducing the idea of scientific progress, Fontenelle did not mean to herald epistemological optimism and trust in the human spirit. On the contrary, the idea of progress compensated for the inadequacy of human thought. Although humans desired knowledge of nature, they had no metaphysical guarantees for the fulfilment of their ambition. The imperfect human spirit

---

[53] Fontenelle 1989–2001, vol. 6, 267; English translation in Fontenelle 1727, 199; cf. Huygens 1888–1950, vol. 21, 210–211.

[54] Fontenelle and Godin 1733, vol. 1, 141; cf. Fontenelle 1702–1742, vol. 7, 15; vol. 8, 125; vol. 9, 127.

[55] Fontenelle 1989–2001, vol. 6, 103; English translation in Fontenelle 1727, 31.

[56] Fontenelle and Godin 1733, vol. 1, 27, the same is repeated in the *éloge* of l'Hôpital, Fontenelle 1989–2001, vol. 6, 99.

[57] Fontenelle 1991, 417; English translation in Fontenelle 1960, 361.

could only neutralize the vastness and inaccessibility of nature by decomposing knowledge into a slow and gradual order of progress that would culminate in a nominalistically conceived set of general principles derived from partial induction rather than metaphysical necessity.

Another relativizing aspect of the necessary order of progress was the reconceptualization of errors in the history of the sciences. Fontenelle regarded reason as unreliable and fundamentally fallible. The second source of knowledge, experience, needed much time for accumulating sufficient amout of findings that could serve as the starting point for inductive reasoning. This epistemological position required a rethinking of the attitude towards ancient philosophers.

In a sense, Fontenelle's standpoint represented the culmination of the transformation in the evaluation of ancient knowledge that took place in the second half of the seventeenth century. Bacon and Descartes were still dismissive of ancient natural philosophy and insisted on its irrelevance to the present. From the limited experience of the ancient philosophers, they inferred the unequivocal superiority of modern philosophers. For example, in his *Dioptrics* (1637), Descartes wrote: 'Carrying our vision much further than our forebears could normally extend their imagination, these telescopes seem to have opened the way for us to attain a knowledge of nature much greater and more perfect than they possessed.'[58]

During the second half of the seventeenth century, the temporal regime of knowledge challenged the radical rejection of past knowledge as it became clear that the empirical horizon and the set of questions posed to nature were changing over time. Discoveries made with the aid of the telescope and microscope demonstrated that the scope of natural philosophy was always relative to the scope of experience, which depended on the power of instruments. According to Pascal, ancient philosophers could not grasp the nature of the Milky Way because they did not have telescopes. Similarly, the Ancients had the right to deny the vacuum, because all their experience led them to such a conclusion. If they had been aware of later experiences with the air-pump, they would probably have changed their minds (Pascal 1998, 457).

Pascal's text signified an important change in the evaluation of classical antiquity. Ancient philosophers 'had the right' to assert their views based on the experiences available to them. They were entitled to defend their opinions, which were obviously wrong from the modern perspective. And not only that, the Ancients should be admired for the knowledge they drew from the few principles they knew. Likewise, they should be excused for their errors because they lacked experience, not 'the force of reason' (Pascal 1998, 457). Pascal's view thus involved historical relativity because it assigned the historical function of a necessary initial phase to error.

Pascal was not alone in these reflections. In his *Machina coelestis,* Hevelius presented similar considerations. In his view, credit was due to the ancient authors for overcoming the initial obstacles (Hevelius 1673, vol. 1, 25). Tournefort remarked that it was necessary to pay respect to old botanists such as Gessner and Dodoens.

---

[58]Descartes 1996, vol. 6, 81; English translation in Descartes 1984–1991, vol. 1, 152.

Since they did not know all the plants they were writing about well, it was not surprising that they were wrong (Tournefort 1698, fol. aiii-v). Tournefort made this remark as a simple matter of fact without charging the Renaissance naturalists with any backwardness.

For many natural philosophers in the second half of the seventeenth century, Greek natural philosophy ceased to be a body of knowledge deserving denial and replacement. It became an initial stage in the process of knowledge that had to be passed through and that deserved respect despite its errors due to lack of experience (Rouvillois 2010, 155). Blumenberg aptly noted that the original act of usurpation of authority found in Bacon and Descartes turned into a historical fatality that sided with the Moderns without stigmatizing the Ancients.[59]

This also meant that true knowledge could not be attained immediately, but only after overcoming the initial phase (or phases) of error. In his notes from 1686, Huygens stated that philosophers arrived at their present level of knowledge because they could resume the works of their predecessors, whose work and diligence should be admired. It was thanks to the knowledge of their predecessors that the present generation would not fall into the same errors as they did: ‘We are allowed to enjoy the fruits of their discoveries without ourselves wandering in the same turning in which they strayed.’[60]

Huygens’s remark suggested that the initial errors of the human spirit had a positive function because they were beneficial to further progress. As at some point in history philosophers rightly considered errors to be true, the distinction between truth and falsity was relativized. The history of human thought, then, was not just the accumulation of knowledge, but also the gradual elimination of error.

Fontenelle echoed these views. The Ancients made many mistakes and errors but that did not mean that they were in any way inferior. ‘In the nature of things it is not given to us to arrive quickly at a reasonable opinion on anything; we must first wander about for a long time and pass through many kinds of mistakes and many kinds of irrelevancies.’[61] This order of progress, according to Fontenelle, was inexorable: ‘it was absolutely necessary to pay to ignorance and error the tribute the Ancients paid.’[62] The relativity of progress thus easily turned into historical fatality.

Individuals contributed to the advancement of the spirit according to their actual capacities. Everyone performed a function that belonged to the moment in the history of the spirit in which the person lived. The Moderns owed the Ancients for having exhausted many errors for them which made it possible to accelerate the advancement of the Moderns to correct knowledge. When we surpassed the first

---

[59] Blumenberg 1986, 185; similarly Mazauric 2007, 237–238.

[60] Huygens 1888–1950, vol. 19, 357; cf. similar words in Pascal 1998, 454–455.

[61] Fontenelle 1991, 418; English translation in Fontenelle 1960, 361; cf. Mazauric 2007, 238–240.

[62] Fontenelle 1991, 418; English translation in Fontenelle 1960, 362; cf. Dagen 1977, 127–128.

discoverer, Fontenelle wrote, he had helped us to do so. 'Thus he still has a share in the glory of our work.'[63]

Relativity of progress naturally applied to the members of the current generation, who helped prepare the path for future discoveries just as the Ancients had. Although the present had surpassed the Greeks, future generations would view the contemporary philosophers as being as blind as the Greeks. Fontenelle, with his usual verbal economy, formulated this relativistic effect by saying that the differences between generations and ages become blurred as time passes. He and his peers would become, in the view of their successors, contemporaries of the Greeks and Romans (Fontenelle 1991, 428–429).

Further relativistic effect concerned knowledge. Fontenelle admitted that knowledge of the present generation represented only a temporary state and would one day be overcome: 'we shall be the ancients someday, and will it not be just for our posterity in its turn to rectify our mistakes and to go beyond us?' (Fontenelle 1991, 421). Fontenelle was consistent in his thinking about the order of progress, and he realized that, unlike the descriptive idea of progress, this mechanism did not make it possible to declare any state of knowledge as final.

The last relativization brought by the Fontenellian order of progress was the insignificance and interchangeability of individuals. According to Seguin, in the *éloges* of individual academicians, Fontenelle presented human knowledge as an essentially historical activity that was continually ongoing. In this structure, each individual was just a link in a chain of knowledge that constituted the essence of history (Seguin 2016a, 416). Similarly, Mazauric remarked that if knowledge followed a logic of autonomous development, then the role played by individuals in this progress was only 'subaltern', since they became mere agents of a process that transcended them (Mazauric 2007, 317).

The insubstantiality of the individual in the history of the human spirit had two aspects. The first was that the individual became only a temporary factor in progress. The present generation could not and would not enjoy the inventions yet to come. Every moment in history was relativized by the future, which promised new conveniences and knowledge that remain denied to those currently living. Gassendi was already aware of this relativity. He realized that the *vita longa* in the form of collaboration over generations meant the end of all hope for the completion of knowledge in the time of one individual life. Definitive knowledge was not reserved for the present generation. Only the descendants would understand nature. The articulation of this idea probably caused bitterness that Gassendi felt necessary to temper with an admonition: we must not envy posterity their happiness (Gassendi 1658, vol. 2, 560a; vol. 4, 504a).

In his *Dialogues des Morts* Fontenelle formulated a very similar reflection in the dispute between Galileo and the Roman cook and gourmet Apicius. Galileo asked how Apicius coped with the idea that new pleasures and inventions would be constantly discovered and yet he was not destined to live until the moment when

[63] Fontenelle 1991, 418; English translation in Fontenelle 1960, 361.

he could take delight in of all the discoveries. And he adds: 'As for new Knowledge, I know you'll never envy its Possessors.' Apicius conciliatorily replied: 'Nature has taken no Care that Men in all Ages shou'd share it [i.e. knowledge] alike.'[64]

Apicius's answer is a reminder of a kind of fatalism that occurs in the Fontenellian concept of the progress of the human spirit. An individual scientist was not promised the chance to enjoy the truth and achievements of scientific progress. Each life span would become a mere episode in the process of the advancement of spirit (Blumenberg 1996, 645–646; Blumenberg 1986, 190). Fontenelle himself occasionally admitted this bitter aspect of the order of progress. In the *HARS* volume for the academic year 1700, he conceded that the individual must make sacrifices in the name of future theories. Empirical practice 'is even more praiseworthy the less brilliant it is, and those who undertake it sacrifice themselves in some way to the glory of those who will make the Systems' (Fontenelle 1702–1742, vol. 2, 3).

Reducing individual research to an episode, Fontenelle's notion of progress also led to the interchangeability of humans. If the progress of the sciences was necessary and subject to its own logic, if each fulfils its function in the advancement of the spirit, then it was possible to proclaim the absolute interchangeability of the Ancients and the Moderns that shocked Fontenelle's contemporaries: 'If we had been in their place we would have done the discovering; if they were in ours, they would add to what has already been discovered. There is no great mystery about that.'[65] This conclusion was a natural consequence of belief in human physiological and cognitive equality. Human brains were always uniform and the same was true for their products. This physicalist approach led to the conclusion that historical cultures differ from each other only in the amount of accumulated knowledge.

As Jauss noted, the initial protest against the norm of *imitation des anciens* led Fontenelle to deny the historical distinctiveness of the Ancients. The idea that historical epochs had their own unmistakable and irreducible uniqueness was foreign to Fontenelle (Jauss 1964, 14–17). This was precisely the point highlighted by the representatives of the Ancients. Unlike the Moderns, who agreed with Fontenelle,[66] the defenders of the Ancients associated the development of knowledge with the unique character of certain cultures, especially classical Greece and Rome. While Fontenelle's theory of progress reduced individuals to episodic tools, the advocates of the Ancients wanted to preserve the importance of individual genius as well as the uniqueness of their cultures. They therefore questioned the uniformity of nature, the irrelevance of cultural differences, and the interchangeability of individualities in scientific progress.

Commenting on his translation of Fontenelle's writings from 1723, the German scholar Johann Christoph Gottsched rejected Fontenelle's physicalist and

[64]Fontenelle 1989–2001, vol. 1, 166; English translation in Fontenelle 1730, 134; cf. Blumenberg 1996, 727.

[65]Fontenelle 1991, 417; English translation in Fontenelle 1960, 360; similar expression can be found in Fontenelle 1989–2001, vol. 3, 235.

[66]See Perrault 1688–1697, vol. 1, 60; Wotton 1694, 9–10; Ayrer 1735, 366–368.

reductionist approach to the quarrel of the Ancients and the Moderns. He did not believe that the uniformity of nature applied to humans. Great artists did not grow like mushrooms from the ground, Gottsched wrote sarcastically. For Gottsched, neither arts nor education were natural processes; they had their own logic, which could not be reduced to natural laws. Artistic achievements were expressions of the soul, not manifestations of brain matter, as Fontenelle believed (Gottsched 1730, 257–258; 259–260). Gottsched therefore doubted that if the Moderns had been in the place of the ancient Greeks, they would have known the same things. Otherwise, the ancient Gauls and Germans would also have known the same as the Greeks (Gottsched 1730, 268). Gottsched was convinced that great discoveries and great cultural achievements had to do not only with the position in the history of the human spirit but with the unmistakable nature of individual cultures. He thus opened a discussion which surpassed the quarrel of the Ancient and the Moderns and has not yet been closed.

6.6 The Acceptance of Fontenelle's Concept by Other Academicians

The history of the human spirit was originally intended to explain the development of human thought, but when Fontenelle began to insist on the necessity and orderliness of the whole process, it turned into a rigid scheme explicating the course of human history. As Fontenelle and other philosophers came to regard the level of knowledge as the defining feature of each epoch, the idea of a history of the spirit could be transformed into a philosophy of history. The Enlightenment began to convey the order of progress to other fields of human activity and eventually to human history. This process has been described by Jean Dagen in his classical monograph (1977).

While the importance of the concept of the progress of the human spirit was undeniable, we may question whether the members of the Académie shared the idea of progress developed by the institution's lifetime secretary. Such a question is difficult to answer unequivocally. As I showed in Chapter Four, academicians conducting empirical research developed a temporal regime of knowledge that embedded scientific inquiry in a specific relationship to the past, present, and future. Such a temporal epistemology undoubtedly co-determined everyday scientific practice and the formulation of specific results: their relationship to the past, their limited validity in the present, and the expectation of further growth, refutation or refinement of the sciences in the future. The temporal regime of knowledge that is the methodological core of the idea of scientific progress is thus very easy to locate in the concrete research of academics. It is much more difficult to find a connection between temporal epistemology and the interpretation of the history of the sciences, which is what Fontenelle accomplished. Academicians only seldom commented at length on the history of the sciences. If they did so, they articulated their opinions in the introductions and prefaces to their own books, where they sought to situate their research within the history of their field. Let us take two examples:

At the beginning of his *Élémens de botanique* (1698), Tournefort presented a history of botany that contained similar features to Fontenelle's theory of the progress of the human spirit. According to Tournefort, botany reached its present state in several stages (*degrez*), which were linked to historical epochs. Tournefort began his interpretation with Greek botany. Giving the name to plants without a method the Greeks (and Romans) did not cultivate botany as a regular science (*science reglée*). Unfortunately, they also understood the discipline itself primarily in terms of its usefulness for medicine (Tournefort 1698, 5–7).

After the Ancients, botany was cultivated by the Arabs, whom Tournefort reproached for blindly following Galen and obscuring the medical applications of plants. After the Arabs came a long period of ignorance and barbarism that lasted until the fifteenth century, when the revival of ancient botany took place. Renaissance authors certainly had good intentions, Tournefort wrote, but they were too dependent on ancient books and neglected empirical research on the plants themselves. It was not until the late sixteenth century that botanists emancipated themselves from the authority of ancient books. They turned to nature itself, and thus put their science on a firmer footing. The culmination of this development was the research at the Académie des sciences (Tournefort 1698, 8–9).

In Tournefort, then, we find a similar historical scheme to that of Fontenelle. Science developed in stages. At the beginning stood the imperfect science of the Ancients, followed by the dark age of barbarism, then came the Renaissance imitating the Ancients, and finally the beginning of methodical empirical research in the sixteenth and seventeenth centuries. The coincidence of opinion between Tournefort and Fontenelle is demonstrable in this case because when Fontenelle reported Tournefort's research to the readers of the *HARS* volumes, he told the story of botany in exactly the same way (Fontenelle 1702–1742, vol. 2, 71–72).

In addition to the general historical paradigm, Tournefort shared with Fontenelle the same attitude towards the older authors. Tournefort did not mock the botanists of the past, such as Gessner. He spoke of them with respect as predecessors. He regarded the imperfections of their research as a natural expression of their stage of development, the existence of which was necessary for later progress. Tournefort even explicitly admitted that he had profited from everything that earlier botanists had written, and that he would have made no progress in botany if the early botanists had not 'trodden the path' for him (Tournefort 1698, 20).

Finally, we find in Tournefort an expectation of further progress that relativized the current state of knowledge. Tournefort did not expect his classification system of plants to remain unchanged. On the contrary, he thought that other plants would undoubtedly be discovered that would necessitate modifications to his system (Tournefort 1698, 24, 38). He thus considered his work to be a work of its time, which botanists would modify according to the results of future research.

What clearly linked Tournefort with Fontenelle was a general idea of the course of the history of science from its Greek beginnings to the research at the Académie.

Both authors agreed on the cumulative nature of science and, at the same time, on the existence of factors that hindered progress: the absence of sufficient data, the absence of a proper method, dependence on authority, overestimation of one's own capabilities. What was missing in Tournefort's conception, however, was an effort to place the development of botany in the larger perspective of the development of all human knowledge. The botanist did not take advantage of the idea that each phase of botany corresponded to an epistemological phase in the development of the human spirit and of humankind as whole. He focused on the history of his field and lacked the philosophical ambition to grasp it more generally.

From time to time, some academicians attempted to implement such philosophical reasoning. Varignon, Fontenelle's close friend, returned to ancient ideas about gravity at the beginning of his *Nouvelles conjectures sur la pesanteur* (1690). He outlined the Aristotelian theory and described it as ridiculous and childish. The ancient theory was based on imagination, which belonged to poetry. Imagination told philosophers that everything was animated and natural phenomena were caused by spirits and deities (Varignon 1690, 2–3). Fontenelle's history of the progress of the human spirit, especially its anthropologically conceived beginnings, seems to have flashed through these words, but again, based on a few sentences, it is impossible to say for sure how much Varignon shared Fontenelle's beliefs. Varignon himself did not develop any conception of the history of the sciences or of mechanics in his books and *mémoires,* as he always concentrated strictly on scientific issues without overlapping with the philosophy of history.

A second example of a more comprehensive text on the history of a discipline was Cassini's treatise *De l'origine et du progrès de l'astronomie* (1693). Cassini first tried to reconstruct the earliest origins of astronomy on the basis of anthropology and economics. Astronomy arose out of necessity, as it was intended to satisfy the needs of humans in the fields of agriculture, navigation and commerce. After outlining the very beginning of human history, however, Cassini entered into biblical history, and unlike most of his contemporaries, he argued that the biblical patriarchs possessed a highly developed astronomy in antediluvian times (Cassini 1730, 5). Cassini believed that there were astronomical reasons for confirming the biblical story of the patriarchs' extensive knowledge (Swerdlow 1990), and apparently thought that this was a way to reconcile the history of the sciences with biblical history.

After the time of the patriarchs, Cassini came to Greek astronomy, which he evaluated positively, but at the same time indirectly referred to Seneca's saying that great works were not perfect at the beginning (Cassini 1730, 18, cf. 24). Unlike Fontenelle and Tournefort, Cassini did not understand the Middle Ages as a time of barbarism, because he assumed an intercultural *translatio studiorum* that allowed the continuity of astronomy to be preserved. Thus, in the Middle Ages, astronomy and geography were developed by the Arabs, so that both disciplines were constantly improving little by little (Cassini 1730, 19). From medieval Arab astronomy, Cassini moved on to overseas discoveries and then to Renaissance France, but admitted that astronomy had not made much progress in the fifteenth century (Cassini 1730, 19). In the sixteenth century, Copernicus did not bring significant changes to astronomy. The real restorer of astronomy (*restaurateur de l'astronomie*) was Tycho Brahe, who based astronomy on long-term observations. Thanks to patient observation with precise instruments, the progress of astronomy accelerated significantly in the last century, and new discoveries were made every day (Cassini 1730, 30).

Reading Cassini's texts reveals that he had a profound knowledge of ancient literature and astronomy, as Fontenelle also claimed (Fontenelle 1702–1742, vol. 4, 65). For Cassini and his colleagues at the Observatoire, ancient and medieval reports of very rare phenomena, such as comets and new stars, or old data were important because of the more precise determination of the parameters of rare and slow phenomena (*Mémoires de l'Académie royale des sciences* 1729–1734, vol. 8, 197, vol. 7, 194). Therefore, astronomers were constantly turning to the past, looking for scraps of information in ancient texts that could be used for perfectioning astronomy. Cassini obviously searched for any surviving data that might serve to improve current knowledge (*Mémoires de l'Académie royale des sciences* 1729–1734, vol. 10, 56–61, 361–372; cf. *Mémoires de mathématique et de physique* 1702–1742, vol. 3, 170–181).

As Cassini and his colleagues of the Observatoire often worked with the ancient sources, they had a natural tendency to grasp the development of their field as a rather slow but continuous progress, which only accelerated considerably in modern times. The belief in the continuity and cumulative nature of astronomical research was thus probably related to the particular epistemic practices of astronomy based on continuous comparison of the data from various historical periods and sources. On the other hand, Cassini's treatise used continuity and cumulativity as a natural explanatory framework for events. Therefore, his account of the history of astronomy and geography was a chronological list of facts that were not connected by any implicit historical conception, or any inevitable order of progress like we saw in Fontenelle. Cassini shared only the basic cumulativist notion of the steady improvement of astronomy, which consisted in increasing the precision of observations through better instruments and methods.

In one of his articles, Cassini noted that in classical antiquity, geography was 'in its infancy, so to speak' (*Mémoires de l'Académie royale des sciences* 1729–1734, vol. 10, 56). The phrase 'so to speak' (*pour ainsi dire*) suggests that Cassini understood the association of historical periods with periods of epistemological development of the individual as an occasional metaphor, not an essential element of a theory of progress, as was the case with Fontenelle. However, he certainly shared Fontenelle's belief in the acceleration of progress in recent decades (Cassini 1730, 30), which made contemporary astronomy incomparably more advanced in relation to the past, without making the modern astronomers superior to their ancient predecessors. Ancient astronomers could not have achieved better results because they lacked modern methods and instruments (*Mémoires de l'Académie royale des sciences* 1729–1734, vol. 10, 56, 371, vol. 7, 716, 730–732).

Cassini also shared Fontenelle's vision of future progress, which he mostly referred to vaguely as the further 'improvement of astronomy' (*Mémoires de l'Académie royale des sciences* 1729–1734, vol. 7, 233, 457, 463, 608; vol. 8, 52). As we have seen above, Cassini was hesitant about the possible completion of astronomical progress. On the other hand, he certainly did not speculate about the cyclical return of epochs of barbarism. He shared Fontenelle's view that the further development of astronomy depended solely on human diligence, not on the metaphysical necessity of historical cycles. The time for the progress of astronomy was human time. Human activity gave direction and acceleration to the development of astronomy, not blind metaphysical necessity or divine providence. In this view he was again very close to Fontenelle.

If we consult the *mémoires* published by academicians in the *HARS* volumes, we only rarely find claims about the history of human thought. The main reason for the absence of an explicit theory of historical progress in the academicians' individual papers was the descriptive nature of these texts. They explained a scientific problem and described a particular attempt to solve it. Individual *mémoires* thus described in detail specific scientific practices: calculations, the circumstances of experiments, the conduct of dissections, or the acquisition of observational data. Naturally, we can easily recognize particular elements of the temporal regime of knowledge, such as the belief in the suspension of judgment, the provisionality of hypotheses, resistance to systems, acceleration, and the belief in the coming of further discoveries and improvements in the sciences in the future (e.g. *Mémoires de mathématique et de physique* 1702–1742, vol. 2, 295; vol. 3, 57, 181, 246). However, only rarely did statements concerning the history of the sciences appear in the descriptions of research as published in the *HARS* volumes.

On the basis of fragmentary and occasional statements, perhaps we can say that academics were in agreement on the assessment of the sciences in classical antiquity. They emphasized the low level of the ancient sciences due to the absence of instruments and methods, the gradual correction of ancient errors, the emancipation from the authority of the Ancients, and the accelerated accumulation of new findings. For example, in relation to plant reproduction, Dodart wrote that the Ancients could not explain the process. Dodart was a proponent of the preformation theory of reproduction and thought that the grains of plants concealed the germs of whole plants. The Ancients, however, could not develop such a theory because they did not have a microscope at their disposal and were limited to writing down commonly known facts. The modern 'physicists' knew that all reproduction was really just the enlargement of finished germs (*Mémoires de mathématique et de physique* 1702–1742, vol. 2, 244–245, cf. vol. 3, 209, 244).

Academicians thus commented relatively regularly on the Greek and Roman sciences, the representatives of which they regarded as their predecessors. However, we observe little mention of the direction and progress of the sciences. Fontenelle apparently considered his task precisely to embed the specific reports of academics in a historical context. To do this, he used the category of the human spirit, which allowed him to interpret separate and specific investigations as part of a larger historical process. Fontenelle represented the work of his colleagues in this way for decades, and it does not appear that any academician protested against Fontenelle's interpretation of research at the Académie. I think it is possible that Fontenelle's theory of the progress of the human spirit formed a kind of implicit conception of the history of the sciences that was more or less accepted and acknowledged by the members of the Académie. Confirming such a thesis, however, would require a much more intensive examination of the *HARS* volumes, which, despite their importance for the understanding of science at the beginning of the Enlightenment, are deeply under-studied and lack systematic analysis.

Bibliography

Armand, Guilhem. 2008–2009. Science et politique chez Fontenelle: une dynamique de l'affranchissement. *Revue Fontenelle* 6-7: 323–335.

———. 2012. De Fontenelle à Voltaire: de la notion de progrès à celle de génération. *Travaux & documents: La Question de la Génération dans les lettres et les arts* 42: 27–36.

Augustine of Hippo. 1981. *Confessionum libri tredecim*. In *Corpus Christianorum series Latina*, ed. Luc Verheijen, vol. 27. Brepols: Turnhout.

Ayrer, Georg H. 1735. *Dissertatio de comparatione eruditionis antiquae ac recentioris*. In Anthony Blackwell, *De praestantia classicorum auctorum*, Lipsiae: Schuster.

Baron, Hans. 1959. The querelle of the ancients and the moderns as a problem for Renaissance scholarship. *Journal of the History of Ideas* 20: 3–22.

Beugnot, Bernard. 1989. Fontenelle: de l'histoire des savants á l'historien des sciences. In *Fontenelle: actes du Colloque tenu à Rouen du 6 au 10 octobre 1987*, ed. Alain Niderst, 487–496. Paris: Presses Universitaires de France.

Bietenholz, Peter G. 1994. *Historia and fabula. Myths and legends in historical thought from classical antiquity to the modern age*. Leiden/New York/Köln: Brill.

Blum, Paul R. 1999. *Vicissitudo* und *memoria*—Geschichtsphilosophie bei Giordano Bruno. In *Wissen und Gesellschaft in Nürnberg um 1500*, ed. Martial Staub and Klaus Anselm Vogel, 105–120. Wiesbaden: Harrassowitz.

———. 2012. *Giordano Bruno: An introduction*. Amsterdam: Rodopi.

Blumenberg, Hans. 1961. Augustins Anteil an des Begriffs der theoretischen Neugierde. *Revue des Ètudes Augustiniennes* 7: 35–70.

———. 1986. *Lebenszeit und Weltzeit*. Frankfurt am Main: Suhrkamp.

———. 1996. *Die Genesis der kopernikanischen Welt*. Frankfurt am Main: Suhrkamp.

Bossuet, Jacques-Bénigne. 1836. *Discourse sur l'histoire universelle*. In *Oeuvres complètes de Bossuet*, vol. 4, 321–570. Besancon: Outhenin-Chalandre.

Bottin, Francesco, Luciano Malusa, Giuseppe Micheli, Giovanni Santinello, and Ilario Tomio. 1993. *Models of the history of philosophy. Volume 1: From its origins in the Renaissance to the historia philosophica*. Dordrecht/Boston/London: Springer.

Brahe, Tycho. 1913–1929. *Tychonis Brahe Dani opera omnia*, ed. John L. Dreyer, 10 vols. Copenhagen: Gyldendal.

Bruno, Giordano. 1977. *The Ash Wednesday supper*. translated by Edward A. Gosselin and Lawrence S. Lerner. Toronto: Univeristy of Toronto Press.

Burke, Peter. 1980. Renaissance, Reformation, Revolution. In *Niedergang. Studien zu einem geschichtlichen Thema*, ed. Reinhart Koselleck and Paul Widmer, 136–147. Stuttgart: Kett-Cotta.

Bury, John B. 1920. *The idea of progress. An inquiry into its origin and growth*. London: Macmillan.

Byrne, James S. 2006. A humanist history of mathematics? Regiomontanus's Padua oration in context. *Journal of the History of Ideas* 67: 41–61.

Callières, François de. 1688. *Histoire poëtique de la guerre nouvellement declarée entre les anciens et les modernes*. Paris: Auboüin/Emery/Cloussier.

Cassini, Jean-Dominique. 1730. De l'origine et du progrès de l'astronomie. In *Mémoires de l'Académie royale des sciences*, vol. 8, 1–52. Paris: Compagnie des Libraires.

Cellarius, Andreas. 1661. *Harmonia macrocosmica*. Amstelodami: Janssonius.

Claussen, Emma. 2018. A sixteenth-century modern? Ancients and moderns in Loys Le Roy's *De la Vicissitude. Early Modern French Studies* 37: 76–92.
Dagen, Jean. 1977. *L'histoire de l'esprit humain dans la pensée française de Fontenelle à Condorcet*. Paris: Klincksieck.
Dechalles, Claude François Milliet. 1690. Tractatus proemialis de progressu matheseos. In Claude François Milliet Dechalles, *Cursus seu mundus mathematicus*, 1–108. Lugduni: Anissonius.
Delvaille, Jules. 1910. *Essai sur l'histoire de l'idée de progrès. Jusqu'a la fin du XVIIIe siècle*. Paris: Félix Alcan.
Demandt, Alexander. 2011. *Philosophie der Geschichte: Von der Antike zur Gegenwart*. Köln: Böhlau.
Descartes, René. 1984–1991. *The philosophical writings*. translated by John Cottingham, Robert Stoothoff, Dugald Murdoch, and Athony Kenny. 3 vols. Cambridge: Cambridge University Press.
———. 1996. *Œuvres de Descartes*, ed. Charles Adam and Paul Tannery, 11 vols., Paris: Vrin.
Dubos, Jean-Baptiste. 1719. *Réflexions critiques sur la poésie et sur la peinture*. 2 vols. Paris: Mariette.
Duris, Pascal. 2016. *Quelle Révolution Scientifique? Les sciences de la vie dans la querelle des Anciens et des Modernes (XVIe-XVIIIe siècles)*. Paris: Hermann.
———. 2017. Les sciences dans la théorie esthétique de l'abbé Du Bos. *Revue d'Histoire litteraire de la France* 117: 895–915.
———. 2022. William Wotton, la *Royal Society* et l'argument scientifique dans la Querelle. In *Anciens et Modernes face aux pouvoirs. L'Eglise, le Roi, les Académies (1687–1750)*, ed. Christelle Bahier-Porter and Delphine Reguig, 197–213. Paris: Honoré Champion.
Ficino, Marsilio. 1576. *Opera omnia*. Basileae: Officina Henricpetrina.
———. 2001–2006. In *Platonic theology*, ed. James Haskins, 6 vols. Cambridge, MA/London: Harvard University Press.
Findlen, Paula. 2002. Historical thought in the Renaissance. In *A Blackwell companion to western historical thought*, ed. Lloyd Kramer and Sarah Maza, 99–119. Oxford: Blackwell.
Fontenelle, Bernard, de. 1702–1742. *Histoire de l'Académie royale des sciences*. 40 vols. Paris: Boudout.
———. 1727. *Lives of the French, Italian and German philosophers, late members of the royal academy of sciences in Paris*. London: Innys.
———. 1730. *Dialogues of the dead*. translated by John Hughes. London: Tonson.
———. 1753. *The history of oracles*. Glasgow: Urie.
———. 1960. A digression on the ancients and the moderns (1688), translated by Donald Schier. In *The continental model. Selected French critical essays of the seventeenth century in English translation*, ed. Scott Elledge and Donald Schier, 358–372. Minneapolis.
——— 1989–2001. Bernard de Fontenelle, *Œuvres complètes*, ed. Alain Niderst, 9 vols. Paris: Fayard.
———. 1989a. De l'origine des fables. In Bernard de Fontenelle, *Œuvres complètes*, ed. Alain Niderst, vol. 3, 187–202. Paris: Fayard.
———. 1989b. Sur l'histoire. In Bernard de Fontenelle, *Œuvres complètes*, ed. Alain Niderst, vol. 3, 169–186. Paris: Fayard.
———. 1991. Digression sur les anciens et les modernes. In Bernard de Fontenelle, *Œuvres complètes*, ed. Alain Niderst, vol. 2, 411–431. Paris: Fayard.
———. 1994. Préface sur l'utilité des mathématiques et de la physique. In Bernard de Fontenelle, *Œuvres complètes*, ed. Alain Niderst, vol. 6, 37–52. Paris: Fayard.
———. 2016. In *Digression sur les anciens et les modernes et autres textes philosophiques*, ed. Sophie Audidière. Paris: Garnier.
Fontenelle, Bernard de, and Louis Godin. 1733. *Histoire de l'Académie Royale des Sciences, depuis son Etablissement en 1666 jusqu'à 1699*, 2 vols. Paris: Martin, Coignard & Guerin.
Gassendi, Pierre. 1658. *Opera omnia*, 6 vols. Lugduni: Anisson.

Gottsched, Johann Ch. 1730. Anmerkungen zur Abhandlung der Frage vom Vorzuge der Alten oder Neuern. In *Herrn Bernhards Fontenelle Gespräche von Mehr als einer Welt*, edited and translated by Johann Ch. Gottsched, 256–294. Leipzig: Breitkopf.

Günther, Horst. 1996. *Le temps de l'histoire. Expérience du monde et catégories temporelles en philosophie de l'histoire, de saint Augustin à Pétrarque, de Dante à Rousseau*. Paris: Éditions de la Maison des sciences de l'homme.

Håkanson, Harald. 2007. Tycho the prophet: History, astrology and the apocalypse in early modern science. In *The word and the world. Biblical exegesis and early modern science*, ed. Kevin Killeen and Peter J. Forshaw, 137–156. New York: Palgrave.

Hanegraaff, Wouter J. 2005. Tradition—Prisca theologia and philosophia perennis. In *Dictionary of gnosis and western esotericism*, ed. Wouter J. Hanegraaff, 1125–1135. Leiden and Boston: Brill.

Hankins, James. 1991. *Plato in the Italian Renaissance*, 2 vols. Leiden/New York/Köln: Brill.

Hevelius, Johannes. 1673. *Machina coelestis*, 2 vols. Gedani: Reiniger.

Hourcade, Philippe. 2003. Jet de plume ou Projet: *Sur l'histoire* de Fontenelle. *Corpus* 44: 17–33.

Huet, Pierre D. 1722. *Huetiana ou Pensées diverses*. Paris: Estienne.

Huygens, Christiaan. 1888–1950. *Œuvres complètes*, 22 vols. La Haye: Nijhoff.

Idel, Moshe. 2002. Prisca theologia in Marsilio Ficino and in some Jewish treatments. In *Marsilio Ficino. His theology, his philosophy, his legacy*, ed. Michael Allen and Valery Rees, 137–158. Leiden/Boston/Cologne: Brill.

Iliffe, Rob. 2000. The masculine birth of time: Temporal frameworks of early modern natural philosophy. *The British Journal for the History of Science* 33: 427–453.

Israel, Jonathan. 2004. Philosophy, history of philosophy, and l'Histoire de l'Esprit Humain: A historiographical question and problem for philosophers. In *Teaching new histories of philosophy*, ed. J.B. Schneewind, 329–343. Princeton: Princeton University Press.

Jardine, Nicolas. 1984. *The birth of history and philosophy of science: Kepler's a defense of Tycho against Ursus*. Cambridge: Cambridge University Press.

Jauss, Hans R. 1964. Ästhetische Normen und geschichtliche Reflexion in der 'Querelle des Anciens et des Modernes'. In Charles Perrault, *Parallèle des Anciens et des modernes en ce qui regarde les arts et les sciences*, 8–64. München: Eidos.

Kehl, Alois, and Henri-Irénée Marrou. 1978. Geschichtsphilosophie. In *Realexikon für Antike und Christentum*, ed. Theodor Klauser, vol. 10, 703–779. Stuttgart: Hiersemann.

Klempt, Adalbert. 1960. *Die Säkularisierung der universalhistorischen Auffassung: zum Wandel des Geschichtsdenkens im 16. und 17. Jahrhundert*. Göttingen: Musterschmidt.

Koselleck, Reinhart. 1975. Fortschritt: IV: Die Ausprägung des neuzeitlichen Fortschrittsbegriffs. In *Geschichtliche Grundbegriffe. Historisches Lexikon zur politisch-sozialen Sprache in Deutschland*, ed. Otto Brunner, Werner Conze, and Reinhart Koselleck, vol. 2, 384–391. Stuttgart: Klett-Cotta.

Kusukawa, Sachiko. 1995. *The transformation of natural philosophy. The case of Philip Melanchton*. Cambridge: Cambridge University Press.

Le Roy, Loys. 1575. *De la Vicissitude ou variété des choses en l'universe*. Paris: Huilier.

Levitin, Dmitri. 2015. *Ancient wisdom in the age of the new science. Histories of philosophy in England, c. 1640–1700*. Cambridge: Cambridge University Press.

Longepierre, Hillaire-Bernard, de. 1687. *Discours sur les anciens*. Paris: Aubouin.

Marchal, Roger. 1997. *Fontenelle à l'aube des lumières*. Paris: Honoré Champion.

Marsak, Leonard Mendes. 1959. Bernard de Fontenelle: The idea of science in the French Enlightenment. *Transactions of the American Philosophical Society* 49: 1–67.

Mazauric, Simone. 2003. Fontenelle et la construction polémique de l'histoire des sciences. *Corpus* 44: 73–97.

———. 2007. *Fontenelle et l'invention de l'histoire des sciences à l'aube des Lumières*. Paris: Fayard.

———. 2008–2009. Fontenelle: Sciences, libertinage et politique. *Revue Fontenelle* 6-7: 269–282.

McKnight, Stephen. 1991. *The modern age and the recovery of ancient wisdom*. Columbia/London: University of Missouri Press.
Meek, Ronald L. 2011. *Social science and the ignoble savage*. Cambridge: Cambridge University Press.
Melanchthon, Philip, and Caspar Peucer. 1844. *Chronicon Carionis*. In *Corpus reformatorum*, ed. Carl G. Bretschneider, vol. 12, 705–1090. Halle: Schwetske.
Mémoires de l'Académie royale des sciences, depuis 1666 jusqu'à 1699. 1729–1734. 11 vols. Paris: Compagnie des Libraires.
Mémoires de mathématique et de physique, tirez des registres de l'Académie royale des sciences. 1702–1742, 40 vols. Paris: Boudot (the same volumes as Fontenelle 1702–1742 but with separate pagination).
Mullet, Isabelle. 2011. *Fontenelle ou la machine perspectiviste*. Paris: Honoré Champion.
———. 2012. Fontenelle et l'Histoire: du fixisme des passions aux progrès de l'esprit humain. *Dix-huitième siècle* 44: 335–347.
Niderst, Alain. 1972. *Fontenelle à la recherche de lui-même (1657–1702)*. Paris: Nizet.
———. 2002. Le monde de Fontenelle. *Études littéraires* 34: 241–248.
Nippel, Wilfried. 1990. *Griechen, Barbaren und Wilde. Alte Geschichte und Sozialanthropologie*. Frankfurt am Main: Fischer.
Nolte, Paul. 2002. Gleichzeitigkeit des Ungleichzeitigen. In *Lexikon Geschichtswissenschaft. Hundert Grundbegriffe*, ed. Stefan Jordan, 134–136. Stuttgart: Reclam.
Pascal, Blaise. 1998. Préface sur le Traité du vide. In Blaise Pascal, *Œuvres complètes*, ed. Michel Le Guern, vol. 1. Paris: Gallimard.
Pasch, Georg. 1700. *De novis inventis*. Lipsiae: Gross.
Pépin, François. 2012. Fontenelle, premier représentant d'une tradition française d'épistémologie historique. *Dix-huitième siècle* 44: 381–398.
Perinetti, Dario. 2006. Philosophical reflection on history. In *The Cambridge history of eighteenth-century philosophy*, ed. Knud Haakonssen, 1107–1140. Cambridge: Cambridge University Press.
Perrault, Charles. 1688–1697. *Parallèle des anciens et des modernes*, 4 vols. Paris: Coignard.
Pomian, Krzystof. 1988. *L'ordre du temps*. Paris: Gallimard.
Poulouin, Claudine. 1998. *Les temps des origines. L'Eden, le Déluge et "les temps reculés". De Pascal à l'Encyclopédie*. Paris: Honoré Champion.
———. 2006. La Préface sur l'utilité des mathématiques et de la physique de Fontenelle ne vise-t-elle qu'à rendre le progrès des sciences visible au 'gros du monde'? *Revue Fontenelle* 4: 61–78.
Rapp, Friedrich. 1992. *Fortschritt. Entwicklung und Sinngehalt einer philosophischen Idee*. Darmstadt: WBG.
Regiomontanus, Johannes. 1834. Oratio de Alfragano et mathematicis disciplinis. In *Corpus reformatorum*, ed. Carl Bretschneider, vol. 11, 531–544. Halle: Schwetske.
Reinhold, Erasmus. 1551. *Prutenicae tabulae coelestium motuum*. Tubingae: Ulricum.
———. 1834. Oratio de Iohanne Regiomontano. In *Corpus reformatorum*, ed. Carl Bretschneider, vol. 11, 817–826. Halle: Schwetske.
Rheticus, Georg Joachim. 1542. *Orationes duae prima de astronomia & geographia altera de physica*. Nurembergae: Petreius.
———. 1968. *Georg Joachim Rheticus 1514–1574. Eine Bio-Bibliographie*, ed. Karl H. Burmeister, 3 vols. Wiesbaden: Preusser.
Richer, Jean. 1729. Observations astronomiques et physiques. In *Mémoires de l'Academie royale des sciences*, vol. 7, 244–270. Paris: Compagnie des Libraires.
Rioux-Beaulne, Mitia. 2011. Ne livrer qu'une partie de son esprit: Fontenelle devant Descartes. *Corpus: Revue de philosophie* 61: 241–262.
———. 2012. Fable et science chez Fontenelle: un problème de partage. *Dix-huitième siècle* 44: 349–364.

———. 2015. Présentation: De *L'Origine des fables* et *Sur l'histoire*. In Bernard de Fontenelle, *Digression sur les Anciens et les Modernes et autres textes philosophiques*, 103–108. Paris: Gallimard.

Rohbeck, Johannes. 2008. Geschichtsphilosophie der Auklärung. In *Grundriss der Geschichte der Philosophie: Die Philosophie des 18. Jahrhunderts 2: Frankreich*, ed. Johannes Rohbeck and Helmut Holzhey. Basel: Schwabe.

Rossi, Paolo. 1970. The idea of scientific progress. In Paolo Rossi, *Philosophy, technology and the arts in the early modern era*, 63–99. London: Harper.

———. 1984. *The dark abyss of time. The history of the earth & the history of nations from Hooke to Vico*. Chicago: The University of Chicago Press.

Rouvillois, Frédéric. 2010. *L'invention du progrès, 1680–1730*. Paris: CNRS.

Schlobach, Jochen. 1980. *Zyklentheorie und Epochenmetaphorik. Studien zur bildlichen Sprache der Geschichtsreflexion in Frankreich von der Renaissance bis zur Frühaufklärung*. München: Fink.

———. 2001. Progress. In *Encyclopedia of the Enlightenment*, ed. Michel Delon, 1100–1105. London/New York: Routledge.

Schröder, Winfried. 1990. Querelle des Anciens et des Modernes. In *Europäische Enzyklopädie zu Philosophie und Wissenschaften*, ed. Hans J. Sandkühler, vol. 3, 983–995. Hamburg: Meiner.

Seguin, Maria Susana. 2001. *Science et religion dans la pensée française du XVIIIe siècle : le mythe du déluge universel*. Paris: Honoré Champion.

———. 2012. Fontenelle et l'histoire de l'académie royale des sciences. *Dix-huitième siècle* 44: 365–379.

———. 2013. La nature dans les écrits de Fontenelle. *Dix-huitième siècle* 45: 97–113.

———. 2016a. Présentation: Éloges des académiciens. In Bernard de Fontenelle, *Digression sur les Anciens et les Modernes et autres textes philosophiques*, ed. Sophie Audidière, 413–418. Paris: Garnier.

———. 2016b. Présentation: Histoire de l'Académie royale des sciences. In Bernard de Fontenelle, *Digression sur les Anciens et les Modernes et autres textes philosophiques*, ed. Sophie Audidière, 319–325. Paris: Garnier.

———. 2017a. Fontenelle, au tournant des XVIIe et XVIIIe siècles. In *Philosophie et libre pensée. Philosophy and Free Thought. XVIIe et XVIIIe siècles*, ed. Lorenzo Bianchi, Nicole Gengoux, and Gianni Paganini, 347–361. Paris: Honoré Champion.

———. 2017b. Merveilles et prodiges dans l'histoire de l'Académie des sciences de Fontenelle. In *Les Sciences et le livre. Formes des écrits scientifiques des débuts de l'imprimé à l'époque moderne*, ed. Joëlle Ducos, 273–285. Paris: Hermann.

———. 2020. Fontenelle, histoire des savoirs et épistemologie historique. *Libertinage et philosophie à l'époque classique (XVIe-XVIIIe siècle)* 17: 247–265.

———. 2022. Anciens et Modernes à l'Académie des sciences. In *Anciens et Modernes face aux pouvoirs. L'Église, le Roi, les Académies (1687–1750)*, ed. Christelle Bahier-Porte and Delphine Reguig, 179–195. Paris: Honoré Champion.

———. 2023. The Royal academy of sciences of Paris and the birth of historical epistemology. In *Ordering knowledge: Disciplinarity and the shaping of European modernity*, ed. Jean-Jacques Chardin, 187–208. Strasbourg: Presses universitaires de Strasbourg.

Seifert, Arno. 1986. Von der heiligen zur philosophischen Geschichte. *Archiv für Kulturgeschichte* 68: 81–117.

Špelda, Daniel. 2013. The search for antediluvian astronomy: Sixteenth- and seventeenth-century astronomers' conceptions of the origin of science. *Journal for the History of Astronomy* 44: 337–362.

———. 2015. From closed cycles to infinite progress: Early modern historiography of astronomy. *History of Science* 53: 209–233.

———. 2018. Huygens' stargazing scientists: The idea of science in *Cosmotheoros*. *History of European Ideas* 44: 1111–1126.

Stadius, Johannes. 1560. *Tabulae Bergenses aequabilis et apparentis motus orbium coelestium*. Coloniae: Brickmann.
Steele, John. 2021. The long history of imaginary inscription: Josephus two pillars in early modern histories of astronomy. In *Afterlives of ancient rock-cut monuments in the near East. Carvings in and out of time*, ed. Jonathan Ben-Dov and Felipe Rojas, 402–406. Leiden: Brill.
Swerdlow, Noel M. 1990. Astronomical chronology and prophecy: Jean-Dominique Cassini's discovery of Josephus's great lunisolar period of the patriarchs. *Journal of the Warburg and Courtald Institutes* 53: 1–13.
Swerdlow, Noel. 1993. Science and humanism in the renaissance: Regiomontanus' Oration on the dignity and utility of the mathematical sciences. In *World changes. Thomas Kuhn and the nature of science*, ed. Paul Horvich, 131–168. Cambridge, MA/London: MIT Press.
Tambrun, Brigitte. 2006. *Pléthon. Le retour de Platon*. Paris: Vrin.
Tournefort, Joseph P. 1698. *Histoire de plantes quie naissent aux environs de Paris*. Paris: L'Imprimerie Royale.
Varignon, Pierre. 1690. *Nouvelles conjectures sur la pesanteur*. Paris: Jombert.
Vasoli, Cesare. 2002. Der Mythos der 'Prisci Theologi' als 'Ideologie' der 'Renovatio'. In *Das Ende des Hermetismus. Historische Kritik und neue Naturphilosophie in der Spätrenaissance*, ed. Martin Mulsow, 19–60. Mohr Siebeck: Tübingen.
Vossius, Gerhard. 1650. *De quattor artibus*. Amstealedami: Blaeu.
Walker, Daniel P. 1972. *The ancient theology. Studies in Christian Platonism from the fifteenth to the eighteenth century*. London: Duckworth.
Werle, Dirk. 2007. *Copia librorum. Problemgeschichte imaginierter Bibliotheken, 1580–1630*. Tübingen: Niemayer.
Wotton, William. 1694. *Reflections upon ancient and modern learning*. London: Buck.
Zedelmaier, Helmut. 2003. *Der Anfang der Geschichte. Studien zur Ursprungsdebatte im 18. Jahrhundert*. Hamburg: Felix Meiner.

Concluding Remarks

In this book, I have attempted to demonstrate that the idea of scientific progress was a methodological means of solving epistemological problems that emerged in early modern natural philosophy. A defining feature of the idea of scientific progress was that it recognized time as a relevant cognitive factor. I therefore sought to find and present the ways that time entered into the knowledge of nature. I wanted to make apparent that such an approach was far from being self-evident in the European philosophical tradition and I therefore outlined epistemological paradigms in which time played no role, in particular the various strands of Renaissance thought.

The introduction of time into the study of nature came about because seventeenth-century natural philosophy understood nature, the universe, and the human relationship to nature as contingent. Metaphysical, cosmological, and epistemological contingency was a condition for the possibility of the idea of scientific progress, and I have devoted considerable attention to its introduction. In the world of anthropocentric teleology, knowledge of nature did not require time; contingent nature could be known neither at once, nor completely, nor immediately. As a result, a new type of natural philosophy emerged that was spread out over time. Knowledge turned into a process taking place over time, and its driving force became curiosity. Natural philosophers could claim this cognitive passion as legitimate precisely in a contingent world in which it was difficult to set limits on human knowledge, the transgression of which could endanger bliss or salvation.

Natural philosophy began to develop a temporal regime of knowledge that represented the epistemological core of the idea of scientific progress. The regime included time as a natural part of the knowledge of nature in several respects that related to the present, the past, and the future. Empirical natural philosophy conducted at Académie royale des sciences did not claim to be definitive and closed. The field presented itself as a set of provisional hypotheses that were likely to change in the future as more data came in. The temporal regime was also concerned with the future: natural philosophers wanted to avoid hastily closing knowledge into a system and saw their task as transmitting their data and hypotheses to the future for later

D. Špelda, *The Origins of the Idea of Scientific Progress*, International Archives of the History of Ideas Archives internationales d'histoire des idées 250,
https://doi.org/10.1007/978-3-031-60526-0

evaluation. As for the past, natural philosophers at the end of the seventeenth century took a different approach from their predecessors at the beginning of the century. They began to respect the previous traditions, which they wanted to neither imitate nor refute. The ancient philosophers became precursors who deserved respect because they had collected data and had helped to prepare the present through their errors. The temporal regime of knowledge thus naturally incorporated the ancient philosophers as predecessors with imperfect knowledge and descendants as successors whose knowledge would necessarily be more perfect. This notion, however, was still deeply counterintuitive to classical scholars in the late seventeenth century.

Natural philosophy, situated between the past, the present, and the future, needed a bearer of knowledge whose duration would correspond to the long time that the knowledge of nature began to require. Nature proved too vast for individuals to explore. At the same time, it was necessary to ensure continuity of research between the past, the present, and the future. The temporal regime thus led to the creation of institutions that transcended the lifetime of individuals and allowed not only for the preservation of data but also for methodology, agendas, and research programmes.

In the preceding chapters, I have shown that the temporal regime of knowledge and the collectivization and institutionalization of research were introduced in Britain and on the continent at least partly based on the same motives. It was not my ambition to provide a systematic comparison of British and French natural philosophy; the references to Boyle, Hooke, Glanvill, and Newton were intended to suggest that these scholars were in many respects in agreement with the representatives of continental natural philosophy, at least as far as the temporal characteristics of natural philosophy were concerned. This was partly due to the fact that natural philosophers came from the same philosophical tradition, the importance of which is rather underestimated in contemporary historical scholarship while the distinctiveness of the British context is somewhat unjustifiably emphasized.

The temporal regime of knowledge methodologically reconfigured the tasks and substance of natural philosophy. The main change was the abandonment of the idea of an eternal and immutable truth, which implied a certain and definite understanding of causes and substances. Contingency brought scepticism about human claims to knowledge of nature, which was compensated by the idea of progress. Instead of aiming at certainty and truth, knowledge became a continual and unfinished process. Such reasoning entailed a fundamental blurring of the boundary between the known and the unknown that became movable and relative. The traditional philosophical (and theological) determination of the limits of knowledge was thus definitively rejected in favour of a conception according to which humans themselves set the limits of their knowledge.

The distribution of knowledge in history was governed neither by Providence, nor by cosmological cycles of flourishing and decay, nor by eschatological inevitability. The main factor of progress was human effort standing against immense nature without any teleological or eschatological support. The idea of scientific progress thus suggested the autonomy of human knowledge. This is also why natural philosophers might have thought that the gain of knowledge was something they could control. Acquiring new knowledge did not result from a metaphysical-historical necessity or the grace of God but from the research itself. This meant that humans could manage the progress of their knowledge and could even accelerate it.

Although the idea of scientific progress was originally an epistemological category, it was transferred to history through the category of the human spirit developed by Fontenelle at the end of the seventeenth century. The idea of the human spirit made it possible to give all human knowledge a unified bearer that was subject to a developmental epistemology, according to which human knowledge as a whole progressed from infantile fables to general laws. The individual epochs of Eurocentrically conceived history were the carriers of the specific phases of this development. The idea of the history of the human spirit also grasped knowledge as the determining factor of human history and the defining feature of individual periods and historical societies. From this point of view, the importance of social and political history was devaluated as a repetitive history of human foolishness and vanity that had no higher meaning.

Fontenelle's outline of the progress of the human spirit always embraced two sides. The first was positive. Its content was a historical depiction of the cumulative gathering of knowledge and inductive discovery of natural laws. The other side was negative because it narrated the story of progress as the diminishing of ignorance. The dynamic factor of the positive side of progress was empirical methods and the accumulation of data. The dynamic factor of negative progress was emancipation from prejudice, errors, and an anthropocentric view of nature. The combination of both the positive and negative sides of progress produced the adventurous story of the emancipation of the human spirit.

When Fontenelle introduced the notion of the progress of the human spirit, it was certainly not in the vein of an optimistic faith in human abilities and exclusiveness. We have seen that the idea of progress was closely linked to scepticism, which concerned the unknowability of nature. What was important was that the prescriptive idea of progress regarded ignorance as transitory and temporary. Thus, the idea of scientific progress was a solution to the inadequacy of human cognitive abilities in relation to the universe. The idea originally expressed neither a sense of triumph nor an ambition for human domination of nature. The idea of progress was an epistemological strategy left to philosophers after the rejection of anthropocentric teleology.

Fontenelle's conception of the progress of the human spirit was not a mere succession of discoveries over time. Therefore, it was fundamentally different from the traditional idea of descriptive progress, which assumed a simple accumulation of discoveries and data over time. As late as the early seventeenth century, philosophers understood progress as a process in which new knowledge was naturally added to the old and humans knew more and more. At the same time, the field of the unknown was being exhausted and becoming smaller and smaller. This idea proved to be an illusion. Especially thanks to optical instruments, natural philosophers became aware of the permanent relativistic mechanism of progress. Natural philosophy broke with the idea that there was a closed set of eternal questions and unchanging problems. New answers raised new questions and unexpected problems, thus giving the achieved progress the ability to produce further progress.

Fontenelle repeatedly emphasized the natural character of acceleration. In his view, the quickening tempo of discoveries was part of the essence of progress. If the sciences were not improving at an ever-faster rate, it would mean that something was holding back their progress. Fontenelle thus could highlight the exclusivity of his epoch. Progressive acceleration elevated the present period above all previous ones. In this way, acceleration was a self-congratulatory category. At the same time, acceleration prevented modern achievements from being lost. By sustaining and accelerating itself, scientific progress achieved sufficient escape velocity to protect the sciences from oblivion or from another period of darkness. As a result, acceleration meant that the progress of the sciences was irreversible.

The history of the human spirit did not proceed haphazardly but had an inner logic and necessity. The discoveries came in succession and according to a developmental epistemology directed towards the inductive derivation of the laws of nature. The connection of epistemological developmental phases with epochs of European history (the Middle Ages) or with nations (the Greeks) made it possible to transfer epistemology to history: the history of human cultures in the Mediterranean area thus suddenly appeared as a gradual expansion of knowledge associated with emancipation from prejudice. In Enlightenment philosophy, the history of the human spirit reduced the history of humanity to the history of knowledge and applied the same rules to it as to the progress of knowledge. The epistemology of progress thus became an explanatory paradigm of human history that replaced both linear eschatology and the cyclical model of history. As a result, the idea of scientific progress went beyond the sphere of epistemology and became the philosophy of history that integrated scientific progress as its crucial component.

At the end of the seventeenth century, it became apparent that if natural philosophy was subordinated to the temporal regime of knowledge, progress started to perpetuate itself, and the time of its conclusion could not be estimated. When knowledge became the defining element of historical epochs, the idea of progress came to exclude contingency (including supernatural interventions) from history. This paradigm allowed the Enlightenment to deny its contingency in history and to declare itself the result and culmination of a long, cumulative, and natural development. Due to this concept of history, the representatives of the Enlightenment acquired the confidence to proclaim the irreversibility of reason and thus quell fears of the coming of a new barbarism. Cosmic contingency was overcome by the realization of a historical meaning whose content was the direction of the human spirit towards the truth.

Index

D. Špelda, *The Origins of the Idea of Scientific Progress*, International Archives of the History of Ideas Archives internationales d'histoire des idées 250,
https://doi.org/10.1007/978-3-031-60526-0

The manufacturer's authorised representative in the EU is Springer Nature Customer Service Centre GmbH, Europaplatz 3, 69115 Heidelberg, Germany. If you have any concerns regarding our products, please contact ProductSafety@springernature.com

Printed and bound by CPI Group (UK) Ltd, Croydon, CR0 4YY
01/10/2026
02225897-0004